ASSESSING MATH CONCEPTS

Kathy Richardson

Assessment Four

NUMBER ARRANGEMENTS

Design, Layout and Editorial Advisor: Janann Roodzant
Layout Assistant: JoEllen Key
Editor: Karen Antell
Illustrations: Linda Starr
Front Cover Design: Lori Young, *Lori I Young Design*
Cover Design/Coordinator: Kenneth W. Harris II, *Mad Monkey Media*

This book is published by Math Perspectives

Math Perspectives Teacher Development Center
P.O. Box 29418
Bellingham, WA 98228-9418
Phone: 360-715-2782

This book is printed on recycled paper.

Distributed by Didax, Inc.

Didax
395 Main Street
Rowley, MA 01969-1207
800-458-0024
www.didax.com

ISBN 978-0-9724238-6-1

DEDICATION

This series of assessments is dedicated to my sister, Janann Roodzant, with more love, gratitude, and respect than I can ever say. And to our parents, Jim and Alice Young, who taught us that in the end, love is all that really matters.

PROFESSIONAL DEVELOPMENT SUPPORT

For information on the Math Perspectives Professional Development Training to support teachers using the *Assessing Math Concepts* assessment series, contact:

Math Perspectives Teacher Development Center
P.O. Box 29418
Bellingham, WA 98228-9419
Phone: 360-715-2782 Fax: 360-715-2783
www.mathperspectives.com

STUDENT INTERVIEW FORMS

To purchase the Assessing Math Concepts Student Interview forms, contact:

Didax, Inc., Distributor
395 Main Street
Rowley, MA 01969-1207
Phone: 1-800-433-4329
www.didax.com

CREDITS

Richardson, Kathy. (1999). *Developing Number Concepts Series: Book 1 Counting, Comparing and Pattern; Book 2 Addition and Subtraction; Book 3 Place Values; Multiplication and Division; Planning Guide.* Parsippany, New Jersey: Pearson Education, Inc., publishing as Dale Seymour Publications.®

TABLE OF CONTENTS

ACKNOWLEDGMENTS

My search to understand how children learn mathematics became a passion for me during my first year of teaching 38 years ago. I remember so well how helpless I felt as a teacher as I looked down at 6-year-old Joey's workbook page to see that he had written a "1" as the answer for every problem on the page. "Joey, what are you doing?" I asked. I will never forget the look on Joey's face when he looked up at me and asked, "Why you always mad to me?" I realized then that I wasn't "mad to" Joey. I could see that he wasn't trying to avoid doing his work. In fact, he was trying to finish an assignment that made no sense to him in the only way he knew how. My frustration was not with him but with my own lack of knowledge. I didn't know why the workbook page that looked so easy to me was so hard for him. I didn't know what to do to help him make sense of it. Thus began my journey to understand how children learn mathematics and my search for ways to help children make sense of the mathematics they are asked to do. My passion for this work has never left me and my search has taken me to many schools throughout the county and allowed me to work with hundreds of children and teachers.

To all those teachers and children who have helped me along the way: thank you for sharing this journey with me.

A special thanks to those teachers who have worked with me over time as I refined the assessments and worked to make the assessment forms more useful. There are too many of you to name but please find yourself in this list and know how grateful I am to you.

Clark County School District , Las Vegas, NV
    MASE Project Teacher Leaders
    MAPS (Math Assessment for Primary Students) Selected Schools involved in the field test of the assessment forms
    Sue Plummer and Lori Squires: Project Leaders

Visalia Unified School District, Visalia, CA
    STEPSS: Math coaches, teacher leaders and classroom teachers

Larabee School, Bellingham, WA
    Cathy Young, math coach

Childs Elementary School, Bloomington, IN
    Chris Oster, teacher

Edmonds School District, Edmonds, Washington
    Edmonds Math Project: Teacher Leaders

There are some people I can never thank enough for all that they have done to help me get this project completed. A special thanks to Janann Roodzant who is the most giving person I know, to Sheryl Russell who is endlessly patient with my struggles to get things right, and to my husband, Tom, for his undying support and encouragement.

And to another very special group of people that make my work possible: A thank you from the bottom of my heart to my friends and colleagues at Math Perspectives: Teacher Development Center. I am grateful every day that you have chosen to be a part of my work and my life.

THE ASSESSMENT SERIES

Assessing Math Concepts

PREFACE

Observing children as they face the challenges of growing and learning is an endlessly fascinating, sometimes puzzling, and always intriguing undertaking. I remember the day many years ago when I decided to accompany my colleague, Margie Gonzales, to the playground for yard duty. While we were discussing some aspect of our ongoing quest to understand how children learn, Anthony, one of her second grade students who had been struggling to learn to read, came running over to her. "Mrs. Gon-gol-es, I know my sight words!" he said with great excitement and pride. Then he proceeded to recite the list of words he had been asked to study, but without benefit of any piece of paper on which the words were written. "Is, of, from, to, on, with...," he said, showing, in spite of his reading difficulty, an ability to learn whatever he believed the school had set out for him–whether it made sense or not. This was a major accomplishment for Anthony. He had worked very hard to memorize that list of words. His time and energy had been spent on something that looked like reading to him but that served him not at all as a reader. This experience with Anthony, who had worked so hard to learn the wrong thing, has come to represent for me all children diligently trying to learn whatever they believe is expected–even when it makes no sense–even when it is the wrong thing.

Anthony came to mind many times in my early years as a teacher, especially when I was teaching mathematics. I observed children working hard to learn the mathematics in front of them. Most of them learned to do the tasks and were able to get right answers, but they seemed to be learning the wrong things. Children who had learned to say the sequence for counting by fives counted objects by moving one at a time instead of five at a time. Children who were able to label the tens and ones in a number counted on their fingers when asked to add 10 to a number. Children who could recite "4 plus 4 is 8" were unable to use that information to solve 4 + 5. Children would carefully follow steps to get answers but were unable to tell if their answers made sense or not. These children, like Anthony, were learning to do what was expected, but what they were learning was not what they needed to know. They could get right answers but did not understand the underlying mathematics with which they were working.

This lack of understanding, so evident in the children with whom I was working, left me with the following questions:

"What are the foundational mathematical ideas children need to know?" and
"How do we know they're learning these important ideas?"

This series of assessments is the result of my search to understand what children need to know to be successful in their study of mathematics and my desire to provide teachers with the tools they need to provide effective and appropriate experiences for their students.

INTRODUCTION TO THE ASSESSMENT SERIES

Assessing Math Concepts is a continuum of assessments that focuses on important core concepts and related "Critical Learning Phases" that must be in place if children are to understand and be successful in mathematics. This assessment series is based on the premise that teachers will be able to provide more effective instruction and ensure maximum learning for each of their students when they are aware of the essential steps that children move through when developing an understanding of foundational mathematical ideas. The data that is gathered and organized using the assessment tools presented here provides teachers with the information that is needed to determine precisely what children need to learn. Students progress confidently when teachers are able to provide appropriately challenging learning experiences for individuals and classroom groups.

The *Number Arrangements Assessment* is fourth in a series of nine books that form the assessment component focusing on numbers to 100. This guide gives the teacher the needed background information for giving and using the assessment tasks, organizing the information, and doing classroom observations. Guidelines for providing appropriate instruction and references to selected resources are included. Copies of these copyrighted Student Interview forms (S.I. forms) are available from the publisher as they are not to be reproduced. Blackline masters of the Classroom Summary sheets (C.S. sheets) used for collecting data and the Assessing at Work forms (A.W. forms) for documenting observations of specific skills while the children are at work in the classrooms are included and may be reproduced for classroom use.

How the Guides to the Assessments Are Organized

Introduction
The guide for each assessment includes an introduction to the assessment series as a whole. The introduction discusses the mathematics that children need to learn and presents an overview of all of the assessments in the series. Information is also provided that helps teachers choose the appropriate assessments for planning classroom instruction and for identifying children who may need some kind of intervention.

The Assessment
Each guide includes the following sections that describe the assessment and how to use it.

I: Learning about the Concept
This section provides the background information that teachers need in order to understand how children develop the concept being assessed. Learning goals related to the concept are listed along with the Critical Learning Phases that children must reach if they are to achieve the goals.

1

II: The Student Interview

An overview of the structure of the Student Interview is presented next to help teachers become familiar with the procedure they are to follow and the questions they are to ask the students. This is followed by examples of actual interviews. The examples help teachers see how the interview unfolds and what kinds of responses they can expect children to make.

Examples of the Student Interview forms and explanations of each section of the interview follow. The actual interview takes just a few minutes to administer. However, there is a great deal of information that can be obtained when teachers know what to look for and how to interpret what they see. Therefore, detailed information about the intent of each of the questions and an explanation of each of the indicators is included.

III: Meeting Instructional Needs

This section presents examples of the Class Summary sheets. These sheets provide a place to organize data so teachers can identify children with similar needs. Also included are guidelines for planning appropriate experiences to meet a range of needs.

IV: Assessing Children at Work

Once the individual interview is given, teachers will be able to observe children's progress over time while they are at work. To assist in this process, an assessment task designed for small groups of students is presented in this section. An example of the Assessing at Work form is included, along with an explanation of the assessment task.

V: Linking Assessment and Instruction

This section describes the instructional needs for each of the indicators on the assessment and refers teachers to the particular activities from the *Developing Number Concept* series to aid those who have access to these resources. Links to other curricula can also be made using the results of these assessment materials.

Appendix of Blackline Masters

Included in this section are blackline masters that individual teachers may reproduce for their own classroom use. They include the Class Summary sheet(s), the Assessing at Work forms, and, when necessary, the masters needed for the Assessing at Work task.

THE MATHEMATICS:
What Do Children Need to Learn?

Mathematics is a complex and multi-faceted discipline that has importance in many areas of human endeavor. It helps us understand the world we live in, from the natural environment to the latest technologies. We use it daily to analyze situations and to solve problems. If the study of mathematics is to be valuable for students, they must learn that mathematics is an important and powerful tool that helps them explore and make sense of the world.

Children's earliest experiences with mathematics build the foundation for the work they will be asked to do in coming years. If children are to be successful in the study of mathematics throughout their schooling, it is vital that the mathematics they learn be meaningful to them. It is only then that they can build on these early experiences.

Far too many children are never given the opportunity to learn that mathematics is a sense-making process. For them, the study of mathematics requires memorizing rules and procedures in order to complete tasks and to get right answers. What they learn turns out to be a set of dead-end skills rather than necessary foundational understandings. In many situations, they are expected to do whatever mathematics has been assigned to their particular grade level whether or not they have the background knowledge they need to understand the mathematics they are asked to do. If they have trouble completing assignments, they are shown how to do the tasks again, and if necessary, they are walked through the required steps. What is missing in this approach is the identification of the foundational ideas necessary for understanding the mathematics inherent in the work the children are expected to do. Rather than repeating a process the child does not understand, it is important to determine what mathematics the child knows and what the child still needs to learn.

What is Basic?
Learning Procedures or the Underlying Mathematics?

There are many children who are successful in completing math assignments but who are not learning the essential mathematics necessary for future success in their study of mathematics. If teachers are to know what they need to do to ensure that children are learning the important foundational ideas, they must understand what mathematics the children are (or are not) learning when completing certain kinds of tasks. They must recognize that what is required of children when they are asked to solve problems using procedures (whether they understand them or not) is different from what is required of children who are asked to demonstrate an understanding of the underlying mathematics.

The following example points out what children need to know in order to get an answer using memorized procedures. This is followed by a description of what children will know if they have learned the mathematics inherent in the problem.

The Problem: 48 + 57

Method One: Getting Answers by Using Procedures

Consider what the children know when they solve the problem using procedures rather than using mathematics.

Step 1 First you have to line up the digits in the correct places:

```
 48
+57
```

Step 2 The digits on the right are labeled "ones." The digits on the left are labeled "tens."

You start with the ones first: 8 + 7 is 15.
(This answer may be arrived at because it has been memorized or more often because children count up from 8 saying 8, 9, 10, 11, 12, 13, 14, 15.)

Step 3 You put down the 5 and carry the 1.

```
 1
 48
+57
  5
```

Step 4 Then you add the tens. 1 plus 4 is 5 and 5 more is 10.

```
 1
 48
+57
105
```

The answer is 105.

When children learn to add, subtract, multiply and divide as a set of procedures and do not also see the underlying logic of the mathematics with which they are working, they may be able to get correct answers, but what they have learned is limited and useful only for solving those particular types of problems. They approach each problem in the same way and do not make connections between one type of problem and another or use what they know to solve similar problems. When children are focused on the procedures rather than the number relationships, they are often unable to tell if their answers are reasonable or not. If they happen to forget a step (such as carrying the 1) they can easily end up with the answer of 915, usually with no concern for whether or not the answer makes numerical sense.

When children learn only to follow procedures without understanding the underlying mathematics, what they are doing is empty of mathematics. Therefore, it is simply not possible for them to build on what has been learned when and if they are asked to deal with more complex ideas. This way of learning mathematics not only limits the ability of students to see relationships and to make connections, but more profoundly, it leads to an attitude towards all mathematics learning which can be summed up by a common comment, "I don't want to know why. Just show me what to do."

On the other hand, when children learn computation as part of the study of important mathematical ideas related to number, they are not only competent when dealing with computational problems, but they also will have learned more mathematics, as we see in the following example.

Method Two: Understanding the Underlying Mathematics
Consider what children know when they have learned the underlying mathematics inherent in the problem   **48 + 57=**

A child who knows the underlying structure of the numbers and their relationships to other numbers might make the following kinds of observations as he or she solves the problem:

> "The answer must be close to 100 because 48 is almost 50 and 57 is a little more than 50.
> Numbers are composed of tens and ones; the 4 is worth 40 and the 5 is worth 50.
> 4 tens and 5 tens make 9 tens. 9 tens is worth 90.
> Since 8 and 2 more makes ten, I can take 2 from the 7 to make a ten.
> 7 less 2 leaves 5. 10 and 5 is 15.
> 90 and 15 make 100 and 5 more. 100 and 5 more is 105."

Or another child might consider the following relationships:

> "48 is 2 less than 50.
> I can take 2 from the 57 to make the 48 into 50.
> 50 and 50 is 100.
> 7 minus the 2 (to make 50) is 5.
> 48 + 57 is 100 and 5 more. That totals 105."

It may take less time in the short run for children to learn the procedures for getting answers than it takes to learn the underlying mathematical concepts. However, simply getting answers is not enough. Learning the essential mathematics is what matters. If the focus is primarily on helping children memorize procedures in order to arrive at answers, children are robbed of important mathematical learning. However, if children are given opportunities to focus on the underlying mathematics, their study of computation can be a critical part of their study of mathematics and can help them develop mathematical thinking and reasoning. They will learn not only how to add, subtract, multiply and divide, but they will also learn number relationships, number composition and decomposition, and the underlying structure of the number system. When children study computation in ways that allow them to understand the inherent mathematics, they are able to work with numbers with understanding and efficiency and are able to build on the mathematics they have learned when studying other areas of mathematics, including fractions, decimals, and percents. Since they are doing mathematics rather than following procedures that are meaningless to them, they are also being prepared for the study of more advanced topics such as algebra and discrete mathematics. Because children's early mathematical experiences have such a profound effect on their future success, teachers must be clear about what mathematics the children should learn in these first years of school and must have ways to determine if the concepts and skills children are learning are meaningful to them.

The Development of Number Concepts

What children know and understand about number and number relationships impacts every other area of mathematical study. Children cannot analyze data, determine functional relationships, compare measures of area and volume or describe relative lengths of sides unless they can use numbers in meaningful ways. Number concepts are the foundation that children must have in order to achieve high standards in mathematics as a whole.

The important mathematics that children must learn to solve problems has been organized for this set of assessments into the following core topics:

Counting

Number Relationships

Number Composition and Decomposition to 20

Place Value: Number Composition and Decomposition of Numbers to 100

Counting
When children first learn to count, they think of numbers as one and another one and another one. I refer to this stage as the "Count and Land" stage. As children develop competence, they will not only count with accuracy and ease in a variety of situations, but they will also develop a sense of the quantity of the numbers they are working with. Their focus will move from knowing what number they landed on to making reasonable estimates and noting the reasonableness of the outcome of their counting.

Beginning Number Relationships
After children learn to count, their next task is to build a sense of quantities and relationships between those quantities. The first relationships children learn are usually "one more" and "one less." Later, children begin to think about the quantity itself in relationship to other quantities and begin to make comparisons between numbers.

Number Composition and Decomposition to 20

Before children are able to decompose numbers, they need to recognize that the smaller numbers are contained in the larger numbers and be able to describe the parts of numbers. That is, they will see that the quantity of 7 is not just a collection of ones, but is composed of 3 and 3 and 1 more. With experience composing and decomposing numbers, they will know the parts of numbers and see how they relate to other numbers so well that they will be able to add and subtract with automaticity. For example, they will know that 3 and 4 must be 7 because 3 and 3 are 6 and 1 more is 7. Or they will know that 3 and 4 are 7 so 7-3 must be 4. They will also know that 8 and 8 is 16, so 8 and 7 must be 15.

Place Value: Thinking of Numbers as Tens and Ones

Once children begin to work with larger numbers, they need to recognize the underlying structure of numbers as tens and ones. To do this, children must be able to count groups of tens as though they were single entities. They must also recognize that when you know the numbers of tens and ones you automatically know the total. When they understand that numbers are composed of tens and ones, they will be able to add 10 and subtract 10 without counting.

Place Value: Addition and Subtraction of Two-Digit Numbers

When children understand numbers as tens and ones and know the parts of numbers to 10, they will able to use what they know to combine numbers by forming all the tens possible and determining the number of leftovers. When subtracting, they can use what they know about numbers to break tens apart and recombine what is left.

Critical Learning Phases: The Key to Effective Instruction

As children develop an understanding of these core topics, there are certain essential ideas that are milestones, or hurdles in their growth of understanding. I have identified these stages of learning as Critical Learning Phases. According to the dictionary, a phase is "A particular moment or stage in a process, especially one at which a significant change or development occurs or a particular condition is reached." The Critical Learning Phase that a child has reached determines the way he or she is able to think with numbers and use numbers to solve problems.

For each major mathematical idea, there are certain understandings that must be in place to ensure that children are not just imitating procedures or saying words they don't really understand. When children receive instruction before they have the foundational ideas necessary to understand the mathematics present in the problems they are asked to solve, the best they can do is memorize the steps for getting right answers to these problems. This creates what I call an "illusion of learning" that breaks down at the point when true understanding becomes necessary for further growth. Children who do not see the underlying logic of the mathematics they are working with learn, over time, to stop looking for meaning and to focus only on procedures.

Sometimes the indicators that reveal whether a child understands or does not understand the mathematics he or she is working with are overlooked because the child appears to know in certain situations and assumptions are made that the children know more about the concept than they do. For example, a child may have learned to line up counters and carefully touch each one and tell how many. However, if the child is not able to hold a number in mind, they will count past that number when asked to get that number of objects. A child may be able to count a group of objects in a variety of settings, but if he is still thinking of each number as distinct from every other number, he may be unable to add to that group of objects to make another group and will build a whole new group instead. A child may have memorized the chant, "6 and 6 is 12", but it can be assumed she doesn't really know the meaning behind what she is saying unless she can use what she knows to solve 6 + 7. A child may be able to say that 3 tens and 4 is 34 but doesn't really understand the underlying structure of these numbers unless he can also tell how many there will be if 10 more were added.

Each of the assessments in the series identifies instructional goals for each concept and the related Critical Learning Phases that are important to the development of those core concepts. An overview of the Critical Learning Phases for Numbers to 100 are listed on the following pages.

THE CRITICAL LEARNING PHASES FOR NUMBERS TO 100

Phase: A particular moment or stage in a process, especially one at which a significant change or development occurs or a particular condition is reached.

The Critical Learning Phase that a child has reached determines the way he or she is able to think with numbers and to use numbers to solve problems.

COUNTING

Counts Objects

- Gets a particular quantity
- Keeps track when counting objects
- Remembers how many after counting
- Counts objects by groups (2s, 5s, and 10s)

NUMBER RELATIONSHIPS

Knows One More/One Less

- Knows one more without counting
- Knows one less without counting

Is Developing a Sense of Quantity/Reasonableness

- Reacts to number estimated while counting
- Adjusts estimate while counting and makes a closer estimate

Relates One Number to Another Number

When changing one quantity to another:
- Tells whether to take some away or get some more
- Adds on or takes away by counting on or removing extras
- Knows (tells) how many to add or take away

Compares Numbers

When comparing two different groups:
- Uses what is known about one amount to determine another:
  Ex. This train has 8. This train has 2 more, so there must be 10.
- Adds or takes away from one group to make it the same as another group
- Tells how many more when groups are lined up
  - when the difference is 1 or 2
  - when the difference is more than 2
- Tells how many more when the groups are not lined up
- Tells how many less when the groups are lined up
  - when the difference is 1
  - when the difference is more than 1

NUMBER COMPOSITION AND DECOMPOSITION TO 20

Recognizes Parts of Numbers to 10 in Arrangements

- Recognizes groups of numbers to 5 in a variety of configurations
- Recognizes and describes the smaller parts contained in larger numbers
- Identifies one or more parts and counts the rest (counting on)
- Combines parts of arrangements by knowing

Knows Number Combinations

- Combines parts using relationships
- Knows doubles
- Uses doubles plus one
- Uses doubles minus one
- Combines parts by knowing

Decomposes Numbers to 10

- Figures out the missing part
- Knows the missing part without figuring out

Recognizes Numbers as One Ten and Some More

- Describes a ten as a single entity even though it is composed of 10 single objects
- Organizes numbers into groups of one ten and leftovers
- Knows 10 plus any number from 1 to 10
- Tells how many needed to make 10
- Tells how many leftovers when removing 10 for numbers from 11 to 20
- Combines quantities by reorganizing into one ten and leftovers
- Subtracts quantities by breaking numbers apart and recombining whatever is left

NUMBERS AS TENS AND ONES

Recognizes Numbers as Tens and Ones

- Forms and counts groups of tens
- Knows total instantly when number of tens and ones is known
- Knows 10 more for any two-digit number
- Knows 10 less for any two-digit number

Combines and Separates Tens and Ones

- Tells how many needed to make the next ten
- Combines numbers by forming new tens and leftovers when necessary
- Breaks apart tens when necessary and reorganizes what is left into tens and leftovers

THE ASSESSMENTS:
How Do We Know They're Learning?

This continuum of assessments follows the stages of children's development of the Critical Learning Phases for the core concepts for numbers to 100. The assessments pinpoint what children know and still need to learn and can be used to document children's growth over time.

Teachers can be misled into thinking that children are making the progress necessary to move on to more complex ideas when they get right answers without understanding the underlying mathematics. Therefore, these assessments are designed to yield more information than whether the children can "get it right or wrong." They determine not only their ability to get right answers but identify the actual mathematics the children know as well.

What young children know and understand can never be fully determined through paper and pencil tasks. Teachers can get much more complete and useful information if they watch and interact with the children while they are doing mathematical tasks. How the children respond indicates what Critical Learning Phase they have reached and reveals their level of understanding. Indicators that describe the range of responses and identify children's instructional needs are listed on each of the assessment forms and explained in detail in each book in the series.

Overview of the Assessments

The assessment described in this book is one of a set of 9. The following charts describe the concepts that are assessed in each book in the series and will help teachers see where the assessment they are using fits into the development of number concepts as a whole.

Topic	Assessments	Focus Questions	Range
Counting	**Concept 1:** *Counting Objects* Student Interview: "Counting Objects" Assessing at Work: "Counting Collections"	**Counting** • Can the children use counting confidently and accurately to find out how many? • Can they count out a particular amount? **One More/One Less** • Do they know one more and one less without counting?	**Task 1:** Counting an unorganized group of objects - from 1 or 2 objects to 32 objects **Task 2:** Counting out from 5 to 18 objects **Task 3:** Telling 1 more and 1 less within the range of numbers from 3 to 22 **Extension:** Telling 1 more and 1 less over the decades and beyond 100

Topic	Assessments	Focus Questions	Range
Number Relationships	**Concept 2:** *Beginning Number Relationships* Student Interview: "Changing Numbers" Assessing at Work: "Fix It"	• Can the children use the relationships between numbers to add or take away the appropriate group of objects to change one number into another? • Do they know particular relationships between numbers?	From adding 3 objects to 2 objects to form a group of 5, to knowing how many to add to 11 to make 15
	Concept 3: *Comparing Numbers* Student Interview: "More/Less Trains" Assessing at Work: "Comparing Yarn" and "Comparing Collections"	• Can the children use what they know about one quantity to determine another? • Can they tell how many more or less one number is than another?	From using one train to determine the length of the other, to determining how many more or less one quantity is than another Settings: Comparing two trains up to 11 cubes long with differences from 1 to 3; Comparing parts of a train (6, 4); Comparing two piles with differences of 3 (6 and 9, 9 and 12)

Topic	Assessments	Focus Questions	Range
Addition and Subtraction to 20	**Concept 4:** *Identifying and Combining Parts* Student Interview: "Number Arrangements" Assessing at Work: "Sorting Arrangement Cards"	• Can the children recognize small groups without counting? • Can they recognize and describe the parts of numbers in arrangements?	Instant recognition of the parts of numbers from 3 to 6 Recognizing and combining parts of numbers to 10
	Concept 5: *Number Combinations* Student Interview: "Combination Trains" Assessing at Work: "Number Combination Cards"	• Can they use what they know about one combination to figure out a related combination? • Can they combine parts to 10 without counting? • Can they combine parts to 20 without counting?	Combining related combinations from 2+1 to 8+8
	Concept 6: *Decomposing Numbers to 10* Student Interview: "Hiding Assessment" Assessing at Work: "Grab Bag Subtraction"	• Can they tell the missing part of a number (subtracting) without having to figure it out?	Identifying the missing parts of numbers from 3 to 10
	Concept 7: *One Ten and Some More* Student Interview: "Ten Frames" Part One: Combining Part Two: Taking Away Assessing at Work: "Can You Make a Ten?" "Can You Break Up a Ten?"	• Can they tell how many are needed to make the next ten? • Can they combine 2 single digit numbers to form a ten and leftovers? • Can they subtract a single digit number from a teen number by breaking up a ten?	**Addition:** From adding 10+8 to 8+7 **Extension:** 18+7 **Subtraction:** From subtracting 17-7 to 13-7 **Extension:** 23-7

Topic	Assessments	Focus Questions	Range
Place Value: Number Composition and Decomposition to 100	**Concept 8:** ***Numbers as Tens and Ones*** Student Interview: "Grouping Tens" Part One: Organizing into Tens Part Two: Conservation and Counting Groups Assessing at Work: "Measuring Shapes"	• Can they tell the total amount of a group of objects if they know the number of tens and ones? • Can they add ten more without counting? • Can they take ten away without counting?	**Task 1:** From determining how many in 3 groups of 10 and 4 ones by counting by ones, to knowing 10 more and 10 less without counting **Extension:** From determining 20 more and 20 less by counting, to knowing without counting Knowing what 7 tens and 12 ones total **Task 2:** From not understanding counting by 2s and 5s to counting by 2s and 5s with ease
	Concept 9: ***Combining and Separating Tens and Ones*** Student Interview: "Two-Digit Addition and Subtraction" Part One: Adding up Tens Part Two: Breaking up Tens Assessing at Work: "Tens and Ones Grid Shapes"	• Can they combine quantities by forming new tens when necessary? • Can they subtract from groups of tens and ones efficiently using relationships?	**Part One:** From solving two-digit addition problems by counting all, to solving problems by combining numbers into tens and ones using their knowledge of the parts of numbers **Part Two:** From solving two-digit subtraction problems by counting all, to solving problems by breaking numbers apart and recombining them using their knowledge of parts of numbers

Assessment Guidelines

There are two ways that information about the child's level of understanding and competence with concepts is gathered; 1) through student interviews and, 2) through assessing children while they are at work.

The Student Interview

The student interviews help teachers become familiar with the stages through which children move as they develop particular concepts. The assessments identify a range of instructional needs and can be repeated over time to document children's growth. They are designed to take a short amount of time while giving the teacher a great deal of information about each child.

Assessing Children at Work

Teachers will benefit greatly from the data gathered from individual interviews as it helps them know what to look for when they watch their students at work. The student interview gives teachers the information and insights into their children's thinking so they can focus right away on how the child approaches a task and can easily interpret what they see. They are then able to respond immediately with appropriate support and challenges. What they observe then adds to the information gained from the interview, making their observations much more productive than they would otherwise be.

Taking Time for the Assessments

It is vitally important that teachers devote the time it takes to assessing the instructional needs of their students. The data gathered through the student interviews is focused on foundational understandings that cannot be obtained in any other way and supplies the information necessary for teachers to provide the most focused and appropriate instruction possible. The assessments have been carefully designed to yield a great deal of information in the least amount of time.

It is reassuring to those who are giving the assessments for the first time to know that the time devoted to doing the assessments will lessen with familiarity with the assessments. Once teachers become familiar with the assessments, they will be able to find where the children are on a continuum of learning and will see it is not always necessary to go through the whole assessment with each child. The assessment is complete as soon as the teacher has reached the point in the assessment that identifies the edge of the child's understanding.

The power inherent in these assessments is that they are simple to administer but what can be learned from them is complex. The explanation of each assessment and description of what can be learned can take many pages but the actual assessment can be done quickly and efficiently.

Teachers have found many different ways to accommodate their need to work with individual students. Sometimes they assess while the other children explore math materials, do quiet work such as reading or drawing, or have an assistant or parent helper read to them. Once the teacher has found a way to get this valuable information about each child, he or she sees that nothing else they might provide for the class is more important to the instructional program as a whole than gathering the information they need to provide appropriate instruction for their students.

Using the Assessment Forms

The assessment forms have been carefully crafted to help teachers pinpoint exactly where a child is on a continuum of understanding. The indicators are included on the form to help teachers know exactly what they are looking for and to minimize the amount of writing that will be necessary. Using the form may seem cumbersome at first, but with practice, teachers will become familiar with what they are looking for, the pattern of responses, and the layout of the form itself. Resist the temptation not to use the form, as that will just delay becoming familiar with the format and will make the assessment process less focused and clear.

Getting Authentic Information from the Student Interview

The essential value of the student interview is the insight teachers can get into the child's thinking, so it is important that the setting be conducive to getting authentic responses from the children. To ensure that the interview yields the most valid information possible, it is important for teachers not to give the children support to do the tasks. The teacher's attitude needs to be one of respectful listening and acceptance of wherever the children are in their understanding of mathematics. The interview should be considered an opportunity to gather information without judgment, and it is important that teachers not make evaluative comments, including praise. The information obtained will then allow teachers to provide their students with appropriate instruction.

Using the Assessments for Classroom Instruction

Because number concepts develop in relatively predictable ways, there are particular concepts and Critical Learning Phases that the majority of children at specific grade levels need to work with. Teachers can use these assessments to determine the instructional needs of the children along a continuum as they develop competence with these particular concepts. The following set of Grade Level charts provide guidance for teachers who are going to use the assessments to plan appropriate instruction for the whole class as they focus on various concepts throughout the year.

PRE-KINDERGARTEN		
Concepts	**Assessment Plans**	**Assessments**
Counting Objects	**Begin the year** by determining the number of objects the children can count. **Reassess** throughout the year, adapting the numbers as the children are able to count larger groups	**Assessment 1: Counting Objects** Begin with smaller numbers: use 12, 7, and 4 as the range of numbers and adjust according to the children's responses.
Beginning Number Relationships	**Later in the year,** if you have children who can count 12 or more objects with ease and accuracy, assess their ability to work with **number relationships.**	**Assessment 2: Changing Numbers** Focus primarily on using smaller numbers as assessed in the Going Back section.
Identifying and Combining Parts	**From the middle to the end of the year,** check to see if any of the children are beginning to **recognize small groups.**	**Assessment 4: Number Arrangements** Focus primarily on the recognition of small groups as assessed in the Going Back section.

KINDERGARTEN		
Concepts	**Assessment Plans**	**Assessments**
Counting Objects	**Begin the year** by determining the largest number of objects the children are able to count. **Reassess** throughout the year to check children's growth in facility with counting.	**Assessment 1:** **Counting Objects**
Beginning Number Relationships	**Once** children can count 20 or more objects with ease and accuracy, determine if they can see relationships between numbers.	**Assessment 2:** **Changing Numbers**
Identifying and Combining Parts	**Around the middle of the year,** check to see which children are able to recognize small groups of numbers and are beginning to see the parts of numbers.	**Assessment 4:** **Number Arrangements**
Decomposing Numbers to 10	**Towards the end of the year,** determine which children are able to decompose numbers. This is a challenging concept for many kindergarten children. There may be some children you want to assess earlier, but most children will not need to be given this assessment until late spring.	**Assessment 6:** **The Hiding Assessment**

FIRST GRADE		
Concepts	**Assessment Plans**	**Assessments**
Counting Objects *Beginning Number Relationships*	**Begin the year** by determining what foundational skills are in place. First check to see whether the children can count objects with ease and accuracy and whether they know one more and one less without counting. Also check to see whether the children are seeing relationships between numbers.	**Assessment 1: Counting Objects** **Assessment 2: Changing Numbers**
Comparing Numbers	**After several weeks,** expand the information you have about the children's understanding of numbers and number relationships by assessing their ability to compare numbers.	**Assessment 3: More/Less Trains**
Identifying and Combining Parts	**As you move to a focus on basic facts,** check the children's ability to identify and describe parts of numbers.	**Assessment 4: Number Arrangements**
Number Combinations *Decomposing Numbers to 10*	**Later,** assess the children's knowledge of number combinations. **Reassess** during the next several weeks to check for progress.	**Assessment 5: Combination Trains** **Assessment 6: The Hiding Assessment**
One Ten and Some More	**Towards the end of the year,** as you begin to work with tens and ones, determine whether the children are able to think of numbers as composed of one ten and some ones.	**Assessment 7: Ten Frames**
Numbers as Tens and Ones	**When you begin work with place value,** determine what children understand about numbers as tens and ones.	**Assessment 8: Grouping Tens**

SECOND GRADE		
Concepts	**Assessment Plan**	**Assessments**
Number Combinations *Decomposing Numbers to 10* *Comparing Numbers*	**Begin the year** by checking the children's knowledge of basic facts and their ability to compare numbers.	**Assessment 5: Combinations Trains** **Assessment 6: The Hiding Assessment** **Assessment 3: More/Less Trains**
Beginning Number Relationships *Identifying and Combining Parts*	**Meanwhile,** if you have children who are not proficient with basic facts or comparing numbers, check their abilities to see number relationships and parts of numbers.	**Assessment 2: Changing Numbers** **Assessment 4: Number Arrangements**
One Ten and Some More	**During the next few weeks,** check the children's ability to work with numbers as one ten and some more.	**Assessment 7: Ten Frames**
Numbers as Tens and Ones	**Later,** when you begin work with place value concepts, check children's understanding of numbers as tens and ones.	**Assessment 8: Grouping Tens**
Combining and Separating Tens and Ones	**Towards the end of the year,** assess the children's ability to combine and separate tens and ones when doing two-digit addition and subtraction.	**Assessment 9: Two-Digit Addition and Subtraction**

THIRD GRADE		
Concepts	**Assessment Plans**	**Assessments**
Use these assessments early in the school year to determine what foundational skills the children bring to third grade.		
Combining and Separating Tens and Ones	**Begin the year** by checking the children's ability to add and subtract two-digit numbers.	**Assessment 9: Two-Digit Addition and Subtraction**
If the children have difficulty making or breaking up tens, use the following assessments until you have determined what the children know, and what they still need to learn.		
Numbers as Tens and Ones	**Check** their understanding of numbers as tens and ones, and their ability to add and subract numbers to 20.	**Assessment 8: Grouping Tens** **Assessment 7: Ten Frames**
Decomposing Numbers	**Also,** assess their knowledge of parts of numbers.	**Assessment 6: The Hiding Assessment**
In a few weeks, reassess to see what progress has been made.		

Using the Assessments for Interventions

This series of assessments can be used to identify the needs of individual students who are having difficulty with the mathematics the rest of the class is working on. When assessing an individual student, the goal is to find the "edge of the child's understanding" in order to provide appropriate instruction that can build on what the child already knows. Often a child needs prerequisite skills to prepare him or her for instruction with the concept the class is working with. The following charts suggest a place to start for each grade level Pre-K through Grade 3. The key is to move between assessments until it has been determined what the child knows and can do and what the child still needs to learn.

Pre-Kindergarten

If the child is having difficulty learning to count objects:

Use **Counting Objects (Assessment 1)**.
Check the child's ability to count a small group of objects. Also check to see if the child can "hand you" a particular number of objects taken from a larger group. Begin with smaller numbers than indicated in the assessment. Begin with up to 8 objects and make the numbers smaller if necessary until the child is successful.

If the child is unable to count 2 or 3 objects, check the child's ability to match objects one-on-one. Suggestions for the kind of activity that will help you assess this idea can be found in *Developing Number Concepts: Book 1 Counting, Comparing and Pattern*. See Activity 1-23: Cover the Dots on page 53, and 1-24: Counting with the Number Shapes on page 54.

Kindergarten

If the child is having difficulty learning to count objects:

Use **Counting Objects (Assessment 1)**.
Begin with smaller numbers than indicated in the assessment. Start with 12 objects and make the numbers smaller if necessary until the child is successful.

If the child is able to count 12 objects sucessfully, ask him or her to count a pile of 21 objects.

If the child is unable to count 3 or 4 objects, check the child's ability to match objects one-on-one. Suggestions for the kind of activity that will help you assess this idea can be found in *Developing Number Concepts: Book 1 Counting, Comparing and Pattern*. See Activity 1-23: Cover the Dots on page 53, and 1-24: Counting with the Number Shapes on page 54.

First Grade

If the child is having difficulty counting and/or comparing numbers:

Use **Changing Numbers (Assessment 2)**.
Check to see if the child recognizes that one number is contained in another number and can change one number to another.

If it becomes evident through the Changing Numbers assessment that the child is not counting with ease and facility, check his or her ability to count.
Use **Counting Objects (Assessment 1)**.

If the child is having difficulty learning addition and subtraction facts:

Find out if the child is able to recognize and describe parts of numbers.
Use **Number Arrangements (Assessment 4)**.

If the child is not able to describe parts of numbers, check his or her understanding of the relationships between numbers by using **Changing Numbers (Assessment 2)**.

If the child is able to describe parts of numbers, check his or her understanding of number combinations using **Combination Trains (Assessment 5)** and **The Hiding Assessment (Assessment 6)**.

Second or Third Grade

If the child is having difficulty learning basic addition and subtraction facts:

Check to see whether the child knows any number combinations without counting. Use **Combination Trains (Assessment 5)** and **The Hiding Assessment (Assessment 6)**.

If the child counts to find the total for most of the combinations or is unable to tell the missing parts of numbers larger than 3 or 4, check the child's ability to recognize and describe parts of numbers and his or her ability to recognize small groups without counting. Use **Number Arrangements (Assessment 4).**

If the child is having difficulty comparing numbers or solving comparative subtraction problems:

Determine his or her level of understanding of comparing by using **More/Less Trains (Assessment 3)**.

If comparing numbers is not easy for the child, also check the child's knowledge of particular relationships between numbers by using **Changing Numbers (Assessment 2).**

If the child is having difficulty understanding place value concepts:

Check the child's ability to think of numbers as composed of tens and ones by using **Grouping Tens (Assessment 8)**.

If necessary, check the child's idea of numbers as one ten and leftovers using the Going Back section of the Numbers as Tens and Ones assessment as well as **Ten Frames (Assessment 7).**

If the child is having difficulty working with addition and subtraction of two-digit numbers:

Check the child's knowledge of number combinations to 10 using **The Hiding Assessment (Assessment 6)**. And also check the child's ability to think of numbers as composed of tens and ones by using **Grouping Tens (Assessment 8)**.

If the child can think of numbers as composed of tens and ones, check his or her ability to add and subtract teen numbers using **Ten Frames (Assessment 7)**.

If the child can make one ten and break apart one ten with ease, check his or her ability to work with several tens and ones using **Two-Digit Addition and Subtraction (Assessment 9).**

NUMBER ARRANGEMENTS

Assessment Four

> # NUMBER ARRANGEMENTS
> # ASSESSMENT

LEARNING TO IDENTIFY PARTS OF NUMBERS

Children can solve addition and subtraction problems to 10 through the action of counting, but counting to get answers will not necessarily lead to the learning of basic facts. What we have come to call basic facts can be looked at as the basic composition and decomposition of numbers to 10. When children know the parts of numbers, they automatically know the basic facts. For example, when children know that 8 can be broken into 4 and 4 or 5 and 3, they also know the relationships expressed as 4 + 4 = 8 and 5 + 3 = 8. Those who know all the parts of a number without hesitation and without counting also know the answers to any problems dealing with those amounts. For example, if a child knows that 8 is made up of 6 and 2, and 5 and 3, and 7 and 1, she will also know the following: "I have 6 toy cars. I need 2 more to make 8." "I have 5 cookies and you have 8. You have 3 more than I do." "I put 8 cubes in the bag. You took out 1. There are 7 left in the bag."

Learning the parts of numbers is not simply a matter of memorizing them. The critical underlying idea that children need to understand is that numbers are not just a collection of single units but are made up of parts. However, children who still think of numbers as one and one more and one more do not think of numbers as composed of parts. When asked to find an answer to an addition or subtraction problem, in essence they treat it as a counting problem. When asked to write answers on worksheets, their attention is usually on completing the assigned task and getting the right answers. Commonly, they count to determine "how many" and then write their answers without paying attention to the particular numbers they end up with. As long as they are only asked to get answers rather than to also describe parts, they will continue to think of numbers as a collection of single units. What is essential to learning the basic facts is learning how to recognize and describe the parts of numbers. The following goals and related Critical Learning Phases focus on what is important for children to learn in order to be successful with identifying and combining parts of numbers. These competencies will be assessed through the tools presented in this guide.

The Goals

CHILDREN NEED TO LEARN
- ♦ To recognize the parts of numbers
- ♦ To combine parts of numbers without counting all

Critical Learning Phases
- ♦ Recognizes groups of numbers to 5 in a variety of configurations
- ♦ Recognizes and describes the smaller parts contained in the larger numbers
- ♦ Identifies one or more parts and counts the rest (counting on)
- ♦ Combines parts of arrangements by knowing

The Challenges of Learning about Numbers and Their Parts

Understanding that numbers are composed of parts may seem obvious to adults, but is not at all obvious to young children who have just recently learned one-to-one counting. We see evidence of this when we ask a child who is still thinking of numbers as one and one more and one more to notice the parts. A child who does not yet see the parts of a number will appear puzzled when asked to look at numbers in this way. For example, Michael was making designs using 7 tiles for each one. One of his designs had 4 tiles across the top and 3 underneath. His teacher, Mrs. Daniel, looked at Michael's work and commented, "Oh, I see you made a design with 4 and 3." Michael quickly responded. "No, I didn't. See, I made 7. 1, 2, 3, 4, 5, 6, 7," he counted, proving to his teacher he indeed had 7 tiles in his design." For Michael, "7" was one group composed of several single objects. 7 was 7—not 3 and 4.

Recognizing small groups of up to 5 objects is a prerequisite to seeing the parts that make up numbers. It stands to reason that children will not be able to see that 7 contains 4 and 3, unless they can recognize the 4 and the 3 when they appear separately. Children must be able to recognize the small groups in a variety of configurations. It is important not to make assumptions about children's ability to recognize parts. My own experience as a preschool teacher taught me this. I had given the children practice in recognizing the arrangements on dot cards using the same dot arrangements found on dice. Many of my students learned to recognize these arrangements very quickly. However, one day, I asked the children to use counters to build what they saw on the cards. To my amazement, I found that many of the children did not use the correct number of counters. Instead they made an X shape to match the shape of the five dots, and they made a "squarish" shape to match the arrangement of the nine dots. I thought I was teaching them the quantities, but they were focused on what the cards looked like.

Once children recognize the small groups when arranged in various ways and can see these groups when they occur as parts of larger numbers, they will be ready to learn the various ways particular numbers can be partitioned and described. Through the process of describing the parts in a variety of ways, they will begin to notice relationships such as the fact that 3 + 3 describes 6, but if you move one of the 3 counters, you end up with 2 and 4. They will also begin to see that 2 + 4 and 4 + 2 describe the same amount. Learning the basic facts then becomes a much more manageable task for them.

The Assessment Tools

Four sections in this guide describe the materials that will help teachers assess children's ability to identify and combine parts of numbers. The first section, "The Student Interview," provides needed information for using the forms for the assessment, "Number Arrangements." The next section, "Meeting Instructional Needs," offers help in organizing this assessment information on the Class Summary sheets and provides guidelines for appropriate instruction. The following section, "Assessing Children at Work," describes how to monitor children's progress while at work using the tasks, "Sorting Arrangement Cards." The last section, "Linking Assessment and Instruction," directs teachers to resources that will help them provide instruction to meet the identified needs.

THE STUDENT INTERVIEW

For teachers to find out whether children can recognize and describe the parts of numbers, they must interact with their young students. It is only by listening to the child's explanations that a teacher can learn which strategies the child uses to find the total and therefore ascertain his or her facility with combining parts. The most efficient way to get this kind of information about a child is through a student interview. Before using the student interview "Number Arrangements," the teacher should become familiar with the structure of the assessment and the indicators used to document students' growth. The indicators specific to each area being assessed identify the particular level of facility that the child has reached. They are organized into categories that describe the kind of instruction the child needs, and thus they help teachers interpret what the child has learned. The category "Ready to Apply" means that a child can already do the particular task and is ready to use this skill in other settings. Another category, "Needs Practice," indicates that a child can do a particular task with some level of effort but still needs more experiences in order to develop facility and consistency. The third category, "Needs Instruction," covers the kinds of responses that show a child is not yet able to do the task.

S.I. THE STUDENT INTERVIEW: NUMBER ARRANGEMENTS

The student interview "Number Arrangements" is designed primarily to determine whether children can identify and combine the parts in various dot arrangements. A series of dot cards is presented, and the children are asked to tell how many dots they see. If the children count each one to find out the number of dots, they are shown cards with groups of up to six dots to see whether they can recognize these small groups when presented separately. If, on the other hand, the children recognize and combine parts, they are shown cards with somewhat larger numbers to determine the level of facility they have reached.

This assessment is designed to identify instructional needs for children working at ability levels ranging from counting groups of 3, to combining parts of numbers up to 10, to explaining that "5 and 5 is 10, so 4 + 5 is 9."

"Number Arrangements"

The following chart is intended to help you become familiar with the structure of the student interview before you work with the actual recording sheet. It outlines the steps you will follow and the questions you will ask. On the next two pages are examples of actual interviews, which will give you an idea of what you might expect when giving this assessment to children. Next is a copy of the actual recording form, including the indicators. A step-by-step explanation of each section of the recording form and indicators follows.

Looking at the Structure of the Student Interview

Show the dot cards as pictured below.	
Show Card 1: Ask: *"How many dots?"* *"How did you find out?"*	Show Card 2: Ask: *"How many dots?"* *"How did you find out?"*

Go to one of the columns below.	
GOING BACK If the child **counts all** to determine the totals, show the cards in the column below.	**GOING ON** If the child **recognizes and combines parts**, show the cards described in the column below.
Ask: *"How many dots?"* *"How did you find out?"*	Ask: *"How many dots?"* *"How did you find out?"*
Ask: *"How many dots?"* *"How did you find out?"*	Ask: *"How many dots?"* *"How did you find out?"*
Ask: *"How many dots?"* *"How did you find out?"*	Ask: *"How many dots?"* *"How did you find out?"*
Ask: *"Can you find any groups that you know on this card?"*	Say: *"See if you can find out how many dots without counting all of them."*

Examples of the Student Interview
(allow 2-8 minutes per child)

The following examples of student interviews give you a picture of what you might expect during the interview and will help make the detailed description found on the following pages more meaningful.

KAYLEE:

Who is Kaylee? She could be a first-grader who is more able to combine parts than many of her classmates. She could be a second-grader who is making good progress in combining parts, or she could be a third-grader who has had a hard time learning basic facts but is now learning to combine numbers using the dot cards.

Mrs. Pease holds up a card with 7 dots on it and asks Kaylee, "How many dots?" Kaylee quickly says, "7." "How did you know so fast?" asks Mrs. Pease. "It's easy," Kaylee says. "There's a 5 like the dice and 2 more so that's 7."

Mrs. Pease holds up a card with a different arrangement of 7 dots. "6 and 1 more is 7," says Kaylee even before Mrs. Pease can ask.

Mrs. Pease then shows Kaylee an arrangement of 8 dots. "That one is just like the 9 on the dominos but one is missing so it's 8."

Mrs. Pease skips the next card and gives Kaylee a chance to try one that may offer a bit of a challenge. "I see that's 9," Kaylee says, "because 5 and 5 is 10 but one is missing so it's 9."

Mrs. Pease shows the last card with 10 dots arranged in a less organized way than the others. "See if you can tell how many dots without counting all of them," she says. "Hmm, this one is a little bit hard. I see a 3 and another 3 and that's 6. And then there's 2 and 2. That's 4. So, 6 . . . and 4. Oh yeah, that's 10." Kaylee has shown Mrs. Pease that she knows the parts of numbers to 10 with ease and is ready to apply this knowledge to more challenging tasks.

CORY:

Who is Cory? Cory could be a preschooler who is showing a particular interest in numbers, a kindergartner who is making good progress for this time of year, or a first-grader who needs many more experiences before he will be able to do what most of his classmates are able to do at this time.

Mr. Alma holds up a card with 7 dots on it and asks Cory, "How many dots?" Cory begins to count, saying, "1, 2, 3, 4, 5, 6, 7. There's 7," he says.

"What about this card?" Mr. Alma asks, showing a different card with 7 dots. Cory counts again and tells Mr. Alma, "That one is 7."

Mr. Alma decides to show Cory the cards listed under the "Going Back" column on the Student Interview form to see whether Cory still counts when the cards have fewer dots. The first card has 4 dots. "That's a 2 and 2 and that's 4," Cory says proudly.

Mr. Alma then shows Cory a card with the same arrangement of 4 as on the previous card plus one more dot, expecting Cory will see this and know quickly that there are 5. But Cory doesn't see the card the way Mr. Alma does. "I see 2 going down and 3 going down. 1, 2, 3, 4, 5. It's a 5," he says.

Next, Mr. Alma shows Cory a card with 3 dots. Before he can ask how many, Cory quickly reports, "3."

The last card in this series has 6 dots, but is arranged in an unusual pattern which can be seen in different ways. "Before you count this one, can you find any groups you know on this card?" Mr. Alma suggests to Cory. Cory responds, "I see a 2 and a 3 and a 1. That's all I see." Mr. Alma asks, "Do you know how many that is all together?" Cory proceeds to count them all. "It's 6," he says. Mr. Alma sees that Cory is just beginning to recognize small groups but still needs to count to find the total number beyond 4.

KRIS:

Who is Kris? She could be a second-grader who has had some difficulties in math but is just discovering that numbers are meaningful, a first-grader who is meeting her teacher's expectations for this time of year, or a kindergartner who is showing a special interest in numbers.

Mrs. Vasser holds up a card with 7 dots on it and asks Kris, "How many dots?" Kris stares at the card for a moment and then says, "7." "How did you find out?" asks Mrs. Vasser. "Well, I saw the 3 in the middle and then 2 on the top and 2 on the bottom. And then I counted them." Kris says. "How did you count them?" asks Mrs. Vasser. "I went 3... 4, 5, 6, 7."

"How many dots do you see on this card," Mrs. Vasser asks, showing another card with 7 dots. Kris looks at the card and then says, "I see 3 and 3 on the outside but then there's 1 more. I think that's 1, 2, 3, 4, 5, 6, 7. It's 7." Mrs. Vassar has enough information to know that Kris is seeing parts but is either counting all or counting on to find the totals. So, she stops the assessment here and excuses Kris to go back to the math stations.

"Number Arrangements"

Name: Date:

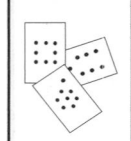

	S.I.
	C.S.
	A.W.

Concept 4: *Identifying and Combining Parts*   *Student Interview*

YOU NEED:
Dot Cards
(BLM 4:3-4)

GOAL:
To determine if the child can recognize parts of a number and combine those parts without having to count all.

PROCEDURE:
Present dot cards as pictured below. Write the child's explanation and circle the appropriate indicator.

Show Card 1	**Identifying and Combining Parts**	Show Card 2	**Identifying and Combining Parts**

Show Card 1

Ask: *"How many dots?"*

[dot card: 5 dots]

"How did you find out?"

Says:

I	Counts all
P-	Identifies parts, but counts all
P	Identifies parts and counts on
A-	Uses related combinations
A	Knows

Show Card 2

Ask: *"How many dots?"*

[dot card: 5 dots]

"How did you find out?"

Says:

I	Counts all
P-	Identifies parts, but counts all
P	Identifies parts and counts on
A-	Uses related combinations
A	Knows

GOING BACK: If the child **counts all** to determine the total, show the cards in the column below. Use the following indicators.

GOING ON: If the child **recognizes and combines parts**, show the cards described in the column below. Use the following indicators.

Recognizing Small Groups	**Identifying and Combining Parts**
INDICATORS	**INDICATORS**

Recognizing Small Groups — INDICATORS

I Counts all	P Quickly combines small groups (2s, 3s)	A- Combines groups of 2s, 3s A Recognizes groups without counting

Identifying and Combining Parts — INDICATORS

I Counts all	P- Identifies parts, but counts all P Identifies parts and counts on	A- Uses related combinations A Knows

Ask: *"How many dots?"* [dot card]	Says:	I	P	A	Ask: *"How many dots?"* [dot card]	Says:	I	P	A
Ask: *"How many dots?"* [dot card]	Says:	I	P	A	Ask: *"How many dots?"* [dot card]	Says:	I	P	A
Ask: *"How many dots?"* [dot card]	Says:	I	P	A	Ask: *"How many dots?"* [dot card]	Says:	I	P	A
Ask: *"Can you find any groups that you know on this card?"* [dot card]	Says:	I	P	A	Say: *"See if you can find out how many dots without counting all of them."* [dot card]	Says:	I	P	A

SUMMARIZING INSTRUCTIONAL NEEDS

Recognizes Small Groups to 5		**Identifies and Combines Parts**	**To 7**	**To 10**
Ready to Apply (A) Recognizes small groups up to 5 without counting		**Ready to Apply (A)** Identifies and combines parts without counting		
Needs Practice (P) Quickly combines small groups (2s, 3s)		**Needs Practice** **(P)** Identifies parts and usually counts on **(P-)** Identifies parts, but usually counts all		
Needs Instruction (I) Counts for most small groups		**Needs Instruction (I)** Doesn't identify parts, counts all		

Understanding the Recording Form and Indicators

Look at the Student Interview form on the adjoining page. In the upper right-hand corner is a picture of several dot cards. This identifies the set of assessment forms for Concept 4: Identifying and Combining Parts. Look for the shaded S.I. box indicating that this is a Student Interview form.

In this assessment, you will present a series of dot cards to the students and ask them to tell you how many dots are on each card. The particular dot cards used in the assessment were selected because the totals can be arrived at in many different ways. Because there are multiple ways that children can answer the question, how they respond is an indication of the level of proficiency they have reached.

✂ SECTION 1:

The first two dot cards to be presented to the children are pictured in this first section of the chart. Show each card to the children and write down what they say. Then circle the appropriate indicator. The indicators are described on the following pages.

Ask: "How many dots?...How did you find out?"

Show Card 1	Identifying and Combining Parts		Show Card 2	Identifying and Combining Parts	
Ask: **"How many dots? ...How did you find out?"** ⠿	Says:	I Counts all P- Identifies parts, but counts all P Identifies parts and counts on A- Uses related combinations A Knows	Ask: **"How many dots? ...How did you find out?"** ⠿	Says:	I- Counts all P- Identifies parts, but counts all P Identifies parts and counts on A- Uses related combinations A Knows

What Can Be Learned From Card One

Children will determine the total number of dots in a great variety of ways. To give you an idea of what you might expect from just this one card, examples of children's responses are presented on the following page.

Card One:

What can be learned from Card One?

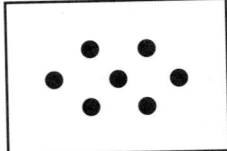

The following examples show how children organize the parts of seven dots. Children's actual words are shown in quotes.

Some children will recognize the 5 dots arranged in the standard dice pattern and will add the 2 on the outside.

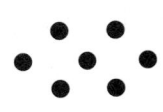 

"I see 5 and 2 more and that's 7."

Some children will count them all.
"I saw 1, 2, 3, 4, 5, 6, 7."

To find out how the child counted, the teacher might ask:
"Did you count like this?"
(starting at the top)

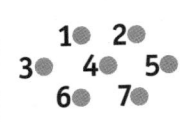

Student responds:
"No, I did this"
(counting in a circle)

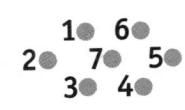

Some children will see the row of 3 in the middle, and 2 dots above and below.

 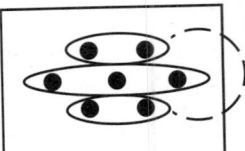

"There's 3 in the middle and then 2 and 2 more."

Some will combine the 2 on top and the 3 in the middle to make 5 and then add 2 more.

 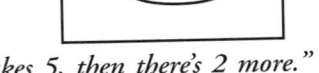

"2 and 3 makes 5, then there's 2 more."

Some will see the 3s and will know that 3+3 is 6 and 1 more makes 7.

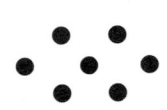 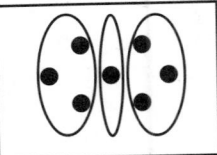

"There's 3 on the sides and 1 in the middle."

Some will combine the 2 and 2 to make 4 and then count up 3 more to make 7.

 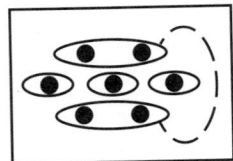

"2 and 2 makes 4 and then I went 5, 6, 7."

Some will see the small groups of 2 and 3 but not report the total.

"I saw 3 going down like that and then 2 and 2."

THE INDICATORS FOR IDENTIFYING AND COMBINING PARTS

When children explain the various ways they arrived at the total, listen for the clues that tell the particular strategy they used. As the examples show, some children count all of the dots, while some describe the parts but still need to count all to find the total. Others recognize one part and then count on. Some children will recognize the parts and know the total number of dots instantly. The following explanation of the indicators will help you interpret children's various responses.

Needs Instruction (I)

(I) Counts all

At this stage, the children do not yet see the parts of a number and must count to determine the total. Children might say: "1, 2, 3, 4, 5, 6, 7. There's 7."

Sometimes children think they are supposed to count even though they might not need to. If you suspect a child is able to see the parts but is not saying so, simply say, "See if you can find any parts that you know without counting."

Sometimes children will say they counted by 2s. If so, ask them to show you how they counted. Some will not actually have counted by 2s, but will have seen groups of 2. They would say something like "There is 2 and 2 and 2." Other children will point to the groups of 2 and count, "2, 4, 6, 8." If any children were indeed able to count by 2s, make a note of that on the form. However, even though counting by 2s is a more advanced way of counting than counting by 1s, it is still a counting strategy and is not evidence that the child is beginning to see parts.

Needs Practice (P)

(P-) Identifies parts, but counts all

At this stage, the children can see the parts of an arrangement and describe them, but they still need to count in order to find out how many all together. This is a sign that the children are beginning to see that numbers are composed of parts. This is sometimes described as "chunking," as it parallels what children are learning in their study of words. A child might say; "I see a 3 and a 4. 1, 2, 3, 4, 5, 6, 7. That makes 7." Another child might count first, saying "1, 2, 3, 4, 5, 6, 7, 8. There is a 4 and a 4 and that makes 8." Watch the children's body language as they figure out the number of dots. Sometimes their words and actions won't match, and their body language will give you some insight into the strategies they are using. Some children will say they "just knew" even though you saw them nodding or moving their eyes to count.

(P) Identifies parts and counts on

Identifying one or more parts in an arrangement and counting on is the first step away from counting all and is important in the child's development of strategies. Children might say: "I saw 4 and then I counted 5, 6, 7," or "I knew there were 3 and then I counted the rest, 4, 5." As important as learning to count on is, it is also a stage in which children get stuck unless they are looking for relationships. If you present ongoing opportunities for children to work with a variety of number arrangements and describe what they see, you will help them to develop even more efficient strategies than counting on.

Ready to Apply (A)

(A-) Uses related combinations

When children begin to use related combinations, it is a sign that they are making connections and developing a network of number relationships. This means they can learn many related facts almost simultaneously rather than learning each combination as a separate fact. Children might say: "I see 3 in the middle and 4 more. 3 and 3 is 6 but the 4 is 1 more so that's 7." Or, "I know 5 and 5 is 10, but that's a 4, so it's 9."

(A) Knows (identifies and combines parts instantly)

The time will come when children just "know" and will no longer need to figure out the totals. The first combination most children learn is 2 + 2 is 4, and soon after that, they will know 2 + 3 is 5. From these beginnings, they will continue to learn more and more number combinations as long as they continue to look for relationships. Children might say, "That one's 5. Because I can tell. See, it's 2 and 3 and that's 5." Or, "There's 5 and 3. I know that one. That's 8."

WHAT CAN BE LEARNED FROM CARD TWO

When you present the second card, take note of whether the children use the same strategy as they did for Card One. Some children realize after they have counted the first card that they don't have to count all the dots. Other children have no other strategies but counting, so, of course, they must count. Card Two can also be seen in more than one way. For example, some children will see the 2 columns of 3 dots on the sides with 1 in the middle. Some will see a row of 3 in the middle with 2 dots above and 2 below. Some will see the 2 and the 3 which is 5, plus the 2 more on the bottom. The indicators used are the same as for Card One.

✂ SECTION 2: Going Back or Going On

After you have asked the child to tell you how many for the first two cards and have marked the appropriate indicators, you will need to decide whether to ask the questions in the column labeled "Going Back," or the questions in the column labeled "Going On."

If the child counted all for both groups and did not identify any parts, show the child the cards in the Going Back column: Recognizing Small Groups to determine whether he or she can recognize any of the small groups when they are presented separately. If the child was able to identify some parts, go on to the cards shown in the "Going On" column, which presents more work on identifying and combining parts.

Ask: "How many dots?" "How did you find out?"

GOING BACK: If the child **counts all** to determine the total, show the cards in the column below.			GOING ON: If the child **recognizes and combines parts**, show the cards described in the column below.		
Recognizing Small Groups			**Identifying and Combining Parts**		
INDICATORS			**INDICATORS**		
I Counts all	P Quickly combines small groups (2s, 3s)	A- Combines groups of 2s, 3s A Recognizes groups without counting	I Counts all	P- Identifies parts, but counts all P Identifies parts and counts on	A- Uses related combinations A Knows
Ask: "*How many dots?*"	Says:	I P A	Ask: "*How many dots?*"	Says:	I P A
Ask: "*How many dots?*"	Says:	I P A	Ask: "*How many dots?*"	Says:	I P A
Ask: "*How many dots?*"	Says:	I P A	Ask: "*How many dots?*"	Says:	I P A
Ask: "*Can you find any groups that you know on this card?*"	Says:	I P A	Say: "*See if you can find out how many dots without counting all of them.*"	Says:	I P A

If you choose Going On, continue as with the first two cards, using the same indicators. If you choose Going Back, use the indicators for Recognizing Small Groups as described below.

INDICATORS FOR RECOGNIZING SMALL GROUPS (GOING BACK)

It is important to find out what base the children have for learning number combinations. Notice whether there are any groups they recognize and whether they are beginning to combine small groups like 2 and 2. Also notice whether they see any relationships between any of the groups. The following are the indicators used to describe the various responses.

Needs Instruction (I)

(I) Counts all
Some children will still need to count for all groups larger than 2.

Needs Practice (P)

(P) Combines groups of 2s, 3s
Some children who are not counting all see that small groups such as 4 and 5 are composed of 2s and 3s but do not yet see groups of 4 and 5 as a whole.

Ready to Apply (A)

(A) Recognizes groups without counting
At this stage, children see the group as a whole and know the number of dots instantly without counting or combining.

The Going Back Cards: Recognizing Small Groups

Do they recognize 4 or do they need to count?

If they know 4, can they instantly see that 1 more is 5, or do they need to count?

Do they recognize 3, or do they need to count?

When showing this card, ask, "Can you find any groups that you know on this card?" (This is a check to see whether the child can see smaller groups as a part of a larger group.)

The Going On Cards: Identifying and Combining Parts

Some children will notice that this looks like the 9 domino pattern with one missing. Others will see the rows of 3 and add the remaining 2.

Children will see this card, like the previous card, in a variety of ways, but this particular arrangement allows you to see whether any child happens to use what they know about 5 + 5 to determine the total.

This card can also be seen in many different ways. Some children will notice the similarity to Card Two as both show 2 rows of 3.

This last card in the series is deliberately designed to determine whether the children can create their own organization even when the card looks less organized than the others. You will find some children reverting to counting all, others seeing parts but still needing to count, and others able to see and combine groups.

✂ SECTION 3: Summarizing Instructional Needs

In this section you will find the chart for summarizing what you have learned from the assessment. When you finish the assessment and excuse the children, take a moment to process what you have learned and record this information on the summary chart. In this chart, you will find two headings under which you can record information: 1) Recognizes Small Groups to 5, and 2) Identifies and Combines Parts to 7 and to 10. The following discussion presents examples that help you see how the summary is designed and what you will want to record. Since it will be helpful later to see the date of the assessment at a glance, let the date serve as the marker in the boxes.

SUMMARIZING INSTRUCTIONAL NEEDS				
Recognizes Small Groups to 5		**Identifies and Combines Parts**	**To 7**	**To 10**
Ready to Apply (A) Recognizes small groups up to 5 without counting		**Ready to Apply (A)** Identifies and combines parts without counting		
Needs Practice (P) Quickly combines small groups (2s, 3s)		**Needs Practice** **(P)** Identifies parts and usually counts on **(P-)** Identifies parts, but usually counts all		
Needs Instruction (I) Counts for most small groups		**Needs Instruction (I)** Doesn't identify parts, counts all		

Learning to Summarize the Student Information

As you process what you learned about the children, consider the following questions: What number arrangements do the children recognize without counting? How do they combine the parts of numbers? What numbers do they know without counting? Which do they figure out by counting on? Which arrangements do they combine by counting all?

The following example of a student interview will help you see how to summarize and record the information you gained from the assessment.

Mr. Chamorro presents the Dot Card 1 to Michael. Michael looks at the card for a moment and says, "I see 3 in the middle. And there's a 2 and a 2." "How many is that all together?" asks Mr. Chamorro. Michael moves his eyes carefully over the card and reports, "7." Mr. Chamorro notes that Michael saw the parts but counted all the dots to determine the total. Mr. Chamorro then shows the Dot Card 2 to Michael. Again Michael looks carefully at the card and says, "I see more 3s. 3 on that side and 3 on that side, and then one in the middle." "How many is that all together?" asks Mr. Chamorro. Michael again looks over the card slowly and says, "I counted 7." Mr. Chamorro notes that Michael counts all to find the total, so he decides to show Michael the cards in the "Going Back" column. He shows the card with 4 dots. Michael says "4" before Mr. Chamorro can even ask "How many?" When he shows the next card, Michael quickly says, "5, 'cause see, 4, 5." He decides to skip the next card because there are only 3 dots on it and Michael has previously shown he recognizes 3. He goes on to the last card, which is arranged in an unusual manner. "Can you see any groups you know without counting?" Mr. Chamorro asks. "I see a 5. See, 'cause there's a 2 and a 3 and that's 5. And another one is 6." Michael says. He concludes that Michael can recognize small groups to 5 and can see parts of larger numbers and now needs practice so he can move past counting all to get the totals. He fills out the summary chart as shown below.

SUMMARIZING INSTRUCTIONAL NEEDS				
Recognizes Small Groups to 5		**Identifies and Combines Parts**	To 7	To 10
Ready to Apply (A) Recognizes small groups up to 5 without counting	3/5	**Ready to Apply (A)** Identifies and combines parts without counting		
Needs Practice (P) Quickly combines small groups (2s, 3s)		**Needs Practice (P)** Identifies parts and usually counts on (P-) Identifies parts, but usually counts all	3/5	
Needs Instruction (I) Counts for most small groups		**Needs Instruction (I)** Doesn't identify parts, counts all		

MEETING INSTRUCTIONAL NEEDS

After you have identified each student's instructional needs, it is important to find practical ways to meet these various needs. The first step is to organize the information you obtained during the individual interview so that common needs are made apparent. The Class Summary sheet for "Number Arrangements" will help you do this. An example of the Class Summary sheet is shown on the adjoining page.

Transfer the individual summary information from each Student Interview form onto the Class Summary sheet. This will give you an overall picture of your class, and you will be able to identify groups of children who have reached similar levels of learning and who therefore share similar instructional needs. A blackline master of the Class Summary sheet can be found in the appendix, BLM 4:1.

C.S. THE CLASS SUMMARY SHEET

In the upper right hand corner of the summary sheet is a coded box that pictures several dot cards. This identifies the set of assessment forms for Concept 4: Identifying and Combining Parts. Look for the shaded C.S. box indicating that this is a Class Summary sheet.

The Class Summary Sheet is intentionally focused on the concept of identifying and combining parts rather than a variety of topics so you will be able to see the instructional needs of your students at a glance.

"Number Arrangements"

Concept 4: Identifying and Combining Parts			Class Summary							S.I. C.S. A.W.

INDICATORS Recognizes Small Groups	INDICATORS Identifies and Combines Parts
I Counts for most small groups **P** Quickly combines small groups (2s, 3s) **A** Recognizes small groups up to 5 (6) without counting	**I** Doesn't identify parts, counts all **P-** Describes parts, but usually counts all **P** Describes parts and usually counts on **A-** Uses related combinations **A** Knows

NAMES	RECOGNIZES SMALL GROUPS			IDENTIFIES AND COMBINES PARTS						COMMENTS
	5			To 7			To 10			
	I	P	A	I	P	A	I	P	A	
Kaylee			3/5			3/5			3/5	
Michael			3/5		P- 3/5					
T.J.			3/6		3/6		3/6			

When you look over the information on the summary sheets, you will see that the children generally fall into three groups. For example, one group of children cannot yet identify parts and will count all the dots in order to determine the total number. Another group of children are able to see the parts in the arrangements but will count on most of the time and know only a few number combinations. Another group of children will use relationships to combine numbers and know many of the combinations. They will need additional challenges.

Grade Level Expectations

Even with the best instruction, children do not all learn at the same time or the same pace. Within every classroom there will be children working at a wide range of levels. As we learn more about how to focus instruction and maximize children's learning, we may find children able to develop understanding and competence faster than we might expect. On the other hand, children who have not had appropriate foundational experiences may not be able to do what we might expect of them under different circumstances. Therefore, the grade level expectations outlined here are intended to be very general guidelines. Although it is useful to know what range is typical, it is the individual achievement of each child that is most important for us to know.

Pre-Kindergarten children will generally count all groups no matter the size, with the possible exception of a group of 2. A few will recognize groups of 3 without counting.

Many kindergarten children will begin the year by counting all groups larger than 2 or 3. By the end of the kindergarten year and the beginning of the first grade year, many children will be able to recognize groups to 5 and will know that 2 and 2 are 4 and 3 and 3 are 6. Some will know that 2 and 3 are 5. Some children will be able to identify small groups when they are a part of larger numbers, but will count all the dots to find the total. A few children will begin to count on from a known group. By the end of first grade, many children will know the parts of numbers to 6 and some parts of numbers to 10. By the end of second grade, with appropriate experiences, most children will be able to combine all dot arrangements to 10 with ease using relationships for any combinations they don't know without counting.

Guidelines for Providing Appropriate Experiences

Identifying parts is an often overlooked but vital step in the development of children's ability to add and subtract. If children are asked to solve problems far beyond what they can solve by focusing on the parts, they will not pay attention to the parts of the numbers they are working with and will be forced to revert to the already learned strategy of counting. All the practice they get will then reinforce their tendency to count rather than help them move to the point when they will not need to count. By the same token, children who work on tasks that do not challenge their thinking or provide new opportunities for learning will not benefit from their work. The challenge is to find practical ways to deal with the range of needs evident on the Class Summary. Teaching to the edge of children's understanding can be realistically managed if you keep some basic ideas in mind.

Children can work side by side with what appears to be the same task and still be working at the level that is appropriate for each one.
It is not always necessary to provide different experiences for children with different needs. In fact, it is not only more practical, it is also more effective to provide tasks that have the potential to be experienced at a variety of levels. The simplest way to meet children's varying needs is to have them work with larger or smaller numbers. For example, children can be working side by side with the same materials. One child can be making designs with 5 toothpicks and another

child can be making designs with 7 toothpicks. One child can be labeling these designs with number combination cards and the other can be writing equations to describe the parts of their designs.

Once you have assessed your students individually, the information you have obtained will help you to efficiently interpret what you see while the children are at work, allowing you to determine quickly what children need.
It takes a long time to gather specific information about children by observing them at work in a classroom setting, but only a few minutes through the student interview. Once you have identified the children's instructional levels in the interview, you will be able to focus on what they are doing while at work and know immediately whether they are working at an appropriate level.

You can meet children's varying needs by interacting with them while they are at work: asking questions, providing support, and posing challenges when appropriate.
The following is an example of how a teacher who knows the individual needs of his students interacts with them during math time.

The children in Mr. Foster's class are busy working with an assortment of materials, making many different arrangements. Some children are creating designs with toothpicks; others are arranging tiles in a variety of ways; still others are building structures with wooden cubes; and some are creating with pattern blocks. The children appear to be working at the same task, but upon closer examination, it is evident that they are working with different numbers. Two children working side by side with the toothpicks are sharing their creations with each other. "Look, I made a house," says Mandy. "It has 2 for the top and 3 on the bottom." "Look at mine. It's a robot. See, the body is 4 and the head is 3 and the legs are 2 and that makes 9," Karl says as he writes the equation on a strip of paper that labels his design. Since Mr. Foster has assessed each one of his children individually, he notices more as he watches his children at work and can more easily tell whether the children are working at an appropriate level. He remembers that Mandy had difficulty recognizing and describing groups when he assessed her a couple of weeks ago, so he is pleased that she is able to describe the parts of 5 as she talks to Karl. Karl, on the other hand, knew the parts of numbers to 8. He is now working to learn the parts of numbers to 9. Both children are getting the practice they need, so Mr. Foster continues to move around the room to observe as many children as he can. He stops by to talk to Tony, who, like Mandy, is working with 5. Tony did not recognize groups and was counting everything larger than 2 when Mr. Foster assessed him. Now he is very carefully and deliberately counting out 5 tiles in order to make a design. Mr. Foster has decided that Tony does not have the prerequisite skills necessary to begin working with parts. Counting is still challenging enough that he wants to make sure Tony can count accurately and confidently before he asks him to work with the parts of numbers. Tony can still work with the same task as the other children, but his focus and learning are completely different from Karl's, for example. Mr. Foster also works with Tony, Natalie, JueMay, and Troy for a few minutes on most days to provide some extra work for recognizing small groups. Mr. Foster walks by Monica. She is counting the parts of her design and writing the equation. Mr. Foster knows she should be able to recognize the parts she is counting and wants to redirect her attention to that fact. He waits until she has another design built and is ready to write an equation and then asks, "Monica, how many toothpicks are on the top of your design?" When Monica begins to count, he stops her and says, "Before you count, do you have an idea of how many are there?" "I see 4 on the top and 4 on the bottom," Monica replies. "So do you need to count, or do you already know?" he asks. She replies, "I guess I already know but I didn't look first." Mr. Foster continues, "I would like you to see what you know without counting every time you write an equation. Don't forget."

Ask questions that will help children think about the concept you want them to learn. For example:

"How many different ways can you make arrangements?"
"Can you tell me about the designs you made?"
"Can you find any designs that are exactly alike?"
"Can you find any of your designs that could be described in the same way?"
"I see 3 and 4 in this design. Can you tell me about this one?"
"Can you tell me about the parts of this number?"
"What if we took the top part away? How many would be left?"
"How many ways can you arrange this number into 2 parts?"
"How do you know you have found all the ways?"
"Can you think of any number combinations you haven't made yet?"
"Can you make another one that looks different from all the others you have made?"

Meet with small groups when particular needs can best be met through teacher-directed activities.
Although many of the children's instructional needs can be met through activities that meet a range of needs, some instructional needs are best met through small group work. Children who are just learning to count will benefit greatly from small group, teacher-directed activities. Children who need practice to develop proficiency and accuracy need more time with independent practice, accompanied by occasional interaction with the teacher about their work.

The following is an example of how a teacher who knows her students' individual needs supports children's learning during a small group activity.

Mrs. Lambert has noticed that the children she has asked to work with her today generally count to tell how many when shown arrangements. She wants to help them see that they don't always have to count. She has picked a set of cards that will help the children use relationships. The first card shows a triangle made from 3 toothpicks. The second card has 2 triangles. The next card also has 2 triangles, but 1 of the triangles has 1 more toothpick along its top. When she holds up the cards, the children put their thumb up when they know how many. She shows the card with 3 toothpicks first. All the children put their thumbs up immediately. "How many toothpicks?" she asks. Robbie says, "3." "See whether this card helps you figure out the next one I am going to show," Mrs. Lambert says. She puts up the card with the 2 triangles. The children's thumbs go up right away but she notices several children bobbing their heads as they count. "How many toothpicks?" she asks. "6" says Alejandro. "Who wants to tell how they figured it out?" Mrs. Lambert asks. She calls on Brittany. "I saw 3 and 3 and that was 6," she says. "How did you know it was 6?" Mrs. Lambert asks. "Well, I saw 3 and another 3. And then I counted," says Brittany." Mrs. Lambert decides not to show the next card with two triangles plus 1 more. Instead, she covers up one triangle and just shows the triangle plus 1. This time she sees the thumbs go up and no heads bobbing. "I could tell it's 4. 'Cause look, 3, 4," Stevie reports. Mrs. Lambert knows the children don't always see the cards the way she expects when she picks them out, and she often has to adjust depending on the children's responses. Even with a small group, there are usually a variety of responses, but she can more easily tailor what she is doing to meet their needs than in the whole group setting.

ASSESSING CHILDREN AT WORK

Observing children's progress as they work is an important part of the daily life of the classroom. So once you have assessed your children through the student interview and have determined their instructional needs, much information about the children's growth in understanding and competence can be gained by observing them over time while they are at work. Sometimes, however, it may be useful to do a more systematic check of their progress and record specific data related to a particular concept for each of the children. The assessment task and recording sheet included here are designed to give you a focused way to keep track of children's ongoing progress identifying and combining parts.

It is recommended that one area be set up as the assessment station where a few children at a time can work with the assessment task while the rest of the children work at other independent station tasks. Over a period of a few days, you will be able to observe all the children you intend to assess.

"Sorting Arrangement Cards"

Concept 4: Identifying and Combining Parts   *Assessing at Work*

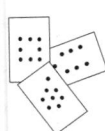

	S.I.
	C.S.
	A.W.

YOU NEED:
Arrangement Cards
(BLM 4:5-10)
Shelf Paper marked in 8"
sections (labeled 3-10)

GOAL:
To determine if the child can identify and combine parts of various number arrangements.

PROCEDURE:
Children pick various Arrangement Cards from a pile, determine how many on each card and place each card in the appropriately labeled section on a large piece of butcher paper. As the children are working, ask them to tell you how they arrived at the total.

WHAT ARE WE LOOKING FOR?

RECOGNIZES SMALL GROUPS	INDICATORS
• Do the children instantly identify small groups up to 5, or do they need to count? • Do they recognize these small groups when they are part of a larger number?	**I** Counts all **P** Combines groups of 2s, 3s **A** Recognizes groups without counting

COMBINES PARTS	INDICATORS
• Do the children find totals for the arrangements? • Do they need to count all or count on, or do they know the totals witout needing to figure them out?	**I** Counts all for most arrangements **P-** Identifies parts, but usually counts all **P** Identifies parts and usually counts on **A-** Uses related combinations (may know some) **A** Knows (identifies and combines parts instantly)

NAMES	RECOGNIZES SMALL GROUPS			IDENTIFIES AND COMBINES PARTS OF 5, 6, 7				IDENTIFIES AND COMBINES PARTS OF 8, 9, 10			
				Combines Parts			Comments	Combines Parts			Comments
	I	P	A	I	P	A		I	P	A	
Sandi			3/22			3/22				3/22	
Joshua		3/22				3/22				3/22	quick at all
Ellie		3/22				A- 3/22	stops to think a bit for 7s		3/22		

A.W. ASSESSING CHILDREN AT WORK: SORTING ARRANGEMENT CARDS

An example of the Assessing Children at Work form for "Sorting Arrangement Cards" is on the adjoining page. In the upper right-hand corner is a picture of several dot cards. This identifies this form as one of a set of assessment forms for Concept 4: Identifying and Combining Parts. The box labeled A.W. is shaded in, indicating that this is the Assessing at Work form. The form includes questions to guide your observations and indicators that help you interpret what you see. The indicators are the same as those used for the student interview, "Number Arrangements."

Run off a copy of the Assessing at Work form for each group you are planning to assess and then record the information that you obtain during the observation on the form as shown. Some teachers may wish to transfer this information to the Class Summary sheet to show the progress the children have made since the Student Interview was given.

Getting to Know the Assessment Task "Sorting Arrangement Cards"

Arrangement Cards are sets of cards picturing a variety of arrangements of dots and sticks. Blackline masters of these cards can be found in the appendix, BLM 4: 5-10. One set of Arrangement Cards includes combinations up to 7 and another set includes combinations up to 10.

When doing the task, "Sorting Arrangement Cards," the children pick various Arrangement Cards from a pile, determine how many objects are on each card selected, and place each card in the appropriate section of a long paper strip. The paper strip is marked off into sections and labeled from 3 to 10. The children will sort as many cards as time allows. You can then observe one child at a time while others in the assessment group are busy doing their work.

Checking for Ongoing Progress

As the children are working, ask them to tell you how they arrived at the total. If children think the point is to get an answer quickly, some will count even when they don't need to. Let them know you want them to see all the groups they can and combine the groups. As they work, notice whether they are able to quickly find the appropriate place to put the cards or whether they have to search. Do they comment on any relationships they see?

Mark the date in the box under the appropriate indicators within the range of numbers you are assessing. The indicators listed on the "Sorting Arrangement Cards" assessment form are the same as those used for the "Number Arrangement" Student Interview.

Children may be performing at different levels depending on the size of the number. For example, you may find that one child is "ready to apply" when recognizing small groups of numbers to 5, "needs practice" when combining numbers to 7, and "needs instruction" when combining numbers to 10. You can use this same assessment after a few weeks and will be able to record children's ongoing progress as they become proficient with identifying and combining parts.

Guidelines for Observing Children

♦ **Make sure children are already familiar with the task.**
It is more efficient if the children are familiar with the task so you are not taking time to teach them how to do the task during this assessment time, but rather are using the task as a tool for determining the stage of their learning.

♦ **Be clear about what you are watching for.**
You can learn many things about children while watching them at work, but to make this kind of assessment manageable, you need to focus specifically on what you need to know to make instructional decisions or to report progress. To guide your observations, refer to the questions included with the assessment form.

♦ **Focus on just one child at a time.**
Don't try to watch too many children at once. Once you have made sure that all the children know what to do and are busy working at the task, you will be more efficient if you focus on one child at a time until you have the information you need.

♦ **Just gather the information. Don't stop and teach.**
The point of the assessment task is to check and record what the children know so far and is not a teaching time. If you just observe and record what the children are able to do, you will be able to watch several children within a short period of time.

LINKING ASSESSMENT AND INSTRUCTION

After you have identified the needs of the children, you will be better able to provide appropriate experiences using whatever instructional materials you have available. The following charts explain the instructional needs of students for each of the indicators on the assessment and refer you to particular activities from the *Developing Number Concepts* series to aid those who have access to these resources. The *Developing Number Concepts* series includes both teacher-directed and independent activities specifically designed to meet the varying needs of students.

There are three books in the series:
   Book 1: Developing Number Concepts: Counting, Comparing and Pattern
   Book 2: Developing Number Concepts: Addition and Subtraction
   Book 3: Developing Number Concepts: Place Value, Multiplication and Division
The activities in each book are coded for easy access. For example, "1:2-23" refers to Book 1, Chapter 2, Activity Number 23.

Providing Appropriate Experiences for Identifying and Combining Parts of Numbers

When you are helping children develop proficiency with identifying and combining parts, it is important to recognize that competency develops over time. Present a variety of activities, allowing children to look for parts of numbers using many different models over several days or even weeks. This will help them make generalizations and integrate the idea that numbers are composed of parts into their thinking about numbers in general. Let their responses dictate the amount of time you spend before moving on to the next level.

Recognizing Small Groups

Before children can learn to identify and combine parts, they need to recognize small groups without counting. Children who always "count all" to determine the number of an arrangement "need instruction" (I) until they can tell how many in a small group without counting. Begin with arrangements that are very small (2 and 3 dots or objects) to help children realize they don't need to count.

Children who recognize small groups of 2s or 3s and use those small groups in identifying groups of 4 or 5 "need practice" (P) until they can recognize groups of 4, 5, and some arrangements of 6 without counting.

Once children recognize groups up to 6 without counting, they are "ready to apply" (A) this ability when learning particular number combinations.

Recommended Activities for Each Indicator

If children need instruction in recognizing small groups:
Provide activities such as those listed below from Developing Number Concepts, Book 2 presenting arrangements of 2, 3, and 4 in a variety of configurations.

NEEDS INSTRUCTION (I) Recognizing Small Groups		
TEACHER DIRECTED ACTIVITIES		**Numbers to 5-6**
2:3-2	Instant Recognition of Number Arrangements	●
2:3-3	Instant Recognition of Number Shapes	●
2:3-4	Instant Recognition of Number Trains	●

If children need practice in recognizing small groups:
Continue to provide activities such as those listed above, including arrangements of 5 and 6 until the children are able to identify the arrangements without counting.

If children are ready to apply their knowledge of small group recognition:
Go on to the activities described in the next section, "Identifying and Combining Parts."

Identifying and Combining Parts

Children who always count without seeing any of the parts in an arrangement "need practice" (P) looking for and describing the parts of numbers.

When children begin to recognize the parts of numbers, they usually still need to count all the objects to determine the total (P-). Eventually they begin to recognize a part and count on (P). Children at these stages need to recognize that they can know the combinations without counting. They need to work first with numbers small enough for them to become aware that they know the totals without counting. Then they can use this insight as they practice combining parts within larger numbers.

Children who use related combinations are building a network of number relationships. This is very important as it makes number combinations much easier to learn. When children can combine parts of numbers to 10 without counting, they are "ready to apply" (A) this knowledge to work with larger numbers.

Recommended Activities for Each Indicator

If children need instruction in identifying and combining parts:
Provide experiences where children describe parts, such as those listed below from Developing Number Concepts, Book 2.

NEEDS INSTRUCTION (I) Identifying and Combining Parts			
TEACHER DIRECTED ACTIVITIES		**Numbers to 7**	**Numbers to 10**
2:2-1	Snap It, Level 1 and Extension	●	●
2:2-2	The Tub Game, Level 1 and Extension	●	●
2:2-3	The Wall Game	●	●
2:2-4	Bulldozer	●	●
2:2-5	The Cave Game	●	●
2:2-7	Finger Combinations	●	●
INDEPENDENT ACTIVITIES		**Numbers to 7**	**Numbers to 10**
2:2-14	Number Arrangements: Using Cubes	●	●
2:2-15	Number Arrangements: Using Color Tiles	●	●
2:2-16	Number Arrangements: Using Toothpicks	●	●
2:2-17	Number Arrangements: Using Collections	●	●
2:2-18	Counting Boards: Making Up Number-Combination Stories	●	●
2:3-24	The Tub-Game Station	●	●
2:3-25	The Snap-It Station	●	●

If children need practice in identifying and combining parts:
Present the same activities as above, but encourage the children to see whether they know the totals without counting. Do Level 2 of those activities that have different levels. In addition, present activities such as the following, found in Developing Number Concepts, Book 2.

NEEDS PRACTICE (P) Identifying and Combining Parts	
TEACHER DIRECTED ACTIVITIES	
2:2-25	How Many Ways?
2:2-27	Building and Rebuilding
2:3-13	Counting Boards: How Many Ways?

If children are ready to apply their knowledge of identifying and combining parts:
Continue to provide practice with a variety of problems and with increasingly larger numbers until they know all the combinations to 10 without figuring them out.

APPENDIX

BLACKLINE MASTERS

"Number Arrangements"

Concept 4: Identifying and Combining Parts　　　　　　　*Class Summary*

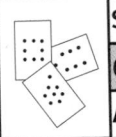

S.I.
C.S.
A.W.

INDICATORS Instant Recognition of Small Groups		INDICATORS Identifies and Combines Parts	
I Counts for most small groups **P** Quickly combines small groups (2s, 3s) **A** Recognizes small groups up to 5 (6) without counting	**I** **P-** **P** **A-** **A**	Doesn't identify parts, counts all Describes parts, but usually counts all Describes parts and usually counts on Uses related combinations Knows	

NAMES	RECOGNIZES SMALL GROUPS			IDENTIFIES AND COMBINES PARTS						COMMENTS
	5			To 7			To 10			
	I	P	A	I	P	A	I	P	A	

"Sorting Arrangement Cards"

Concept 4: Identifying and Combining Parts

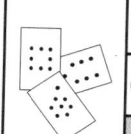

	S.I.
Assessing at Work	C.S.
	A.W.

YOU NEED:
Arrangement Cards
(BLM 4:5-10)
Shelf Paper marked in 8"
sections (labeled 3-10)

GOAL:
To determine if the child can identify and combine parts of various number arrangements.

PROCEDURE:
Children pick various Arrangement Cards from a pile, determine how many on each card and place each card in the appropriately labeled section on a large piece of butcher paper. As the children are working, ask them to tell you how they arrived at the total.

WHAT ARE WE LOOKING FOR?

RECOGNIZES SMALL GROUPS

- Do the children instantly identify small groups up to 5, or do they need to count?
- Do they recognize these small groups when they are part of a larger number?

INDICATORS

I	Counts all
P	Combines groups of 2s, 3s
A	Recognizes groups without counting

COMBINES PARTS

- Do the children find totals for the arrangements?
- Do they need to count all or count on, or do they know the totals witout needing to figure them out?

INDICATORS

I	Counts all for most arrangements
P-	Identifies parts, but usually counts all
P	Identifies parts and usually counts on
A-	Uses related combinations (may know some)
A	Knows (identifies and combines parts instantly)

NAMES	RECOGNIZES SMALL GROUPS			IDENTIFIES AND COMBINES PARTS OF 5, 6, 7				IDENTIFIES AND COMBINES PARTS OF 8, 9, 10			
				Combines Parts			Comments	Combines Parts			Comments
	I	P	A	I	P	A		I	P	A	

Dot Cards for Student Interview "Number Arrangements"

On - 2

Card 1

On - 3

Card - 2

On - 4

On - 1

BLM 4: 3

Copy onto card stock, cut on the dotted lines, and store complete set in a zip bag.

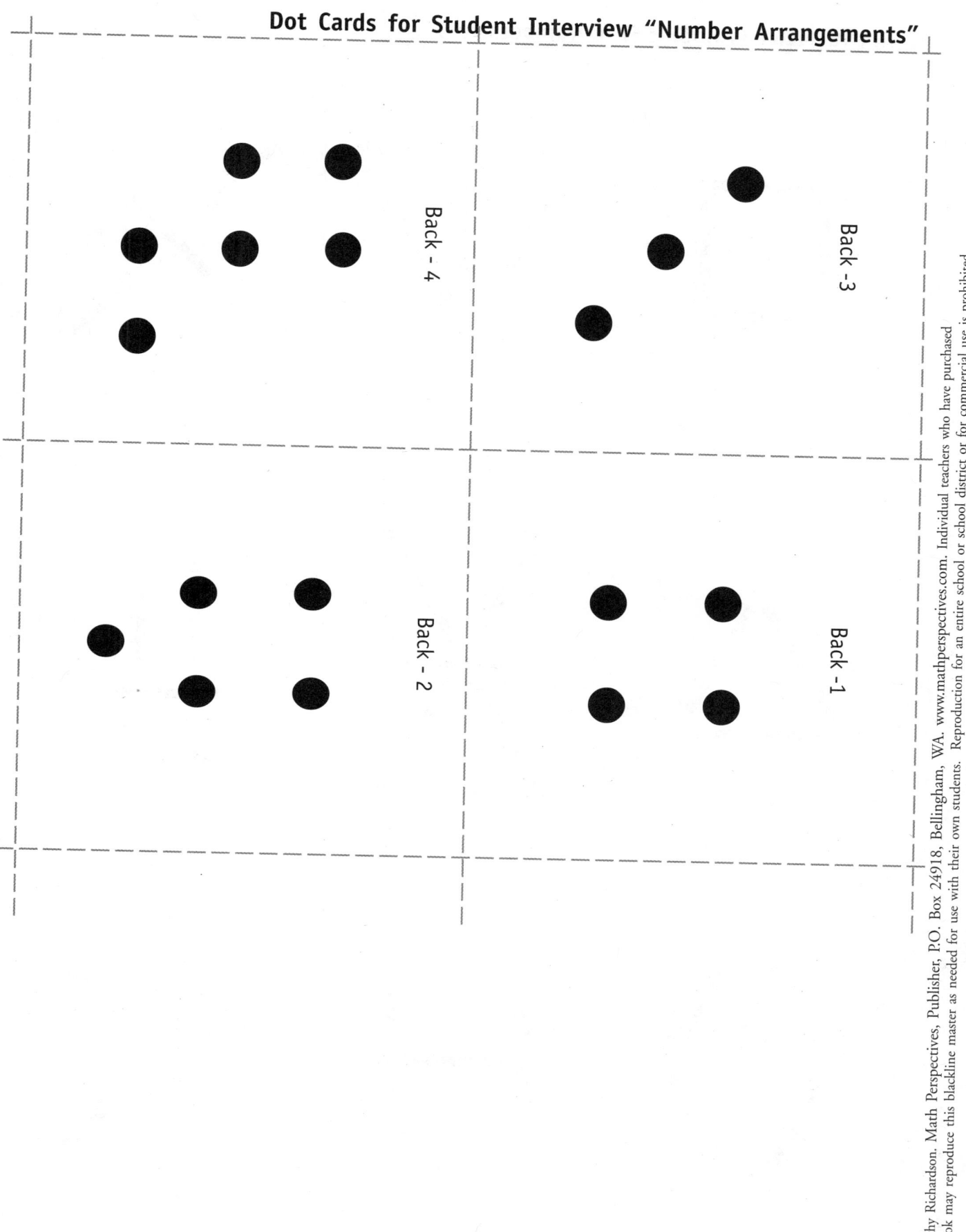

Back - 4

Back - 3

Back - 2

Back - 1

Copy onto card stock, cut on the dotted lines, and store complete set in a zip bag.

ARRANGEMENT CARDS: SET 1 Numbers to 7

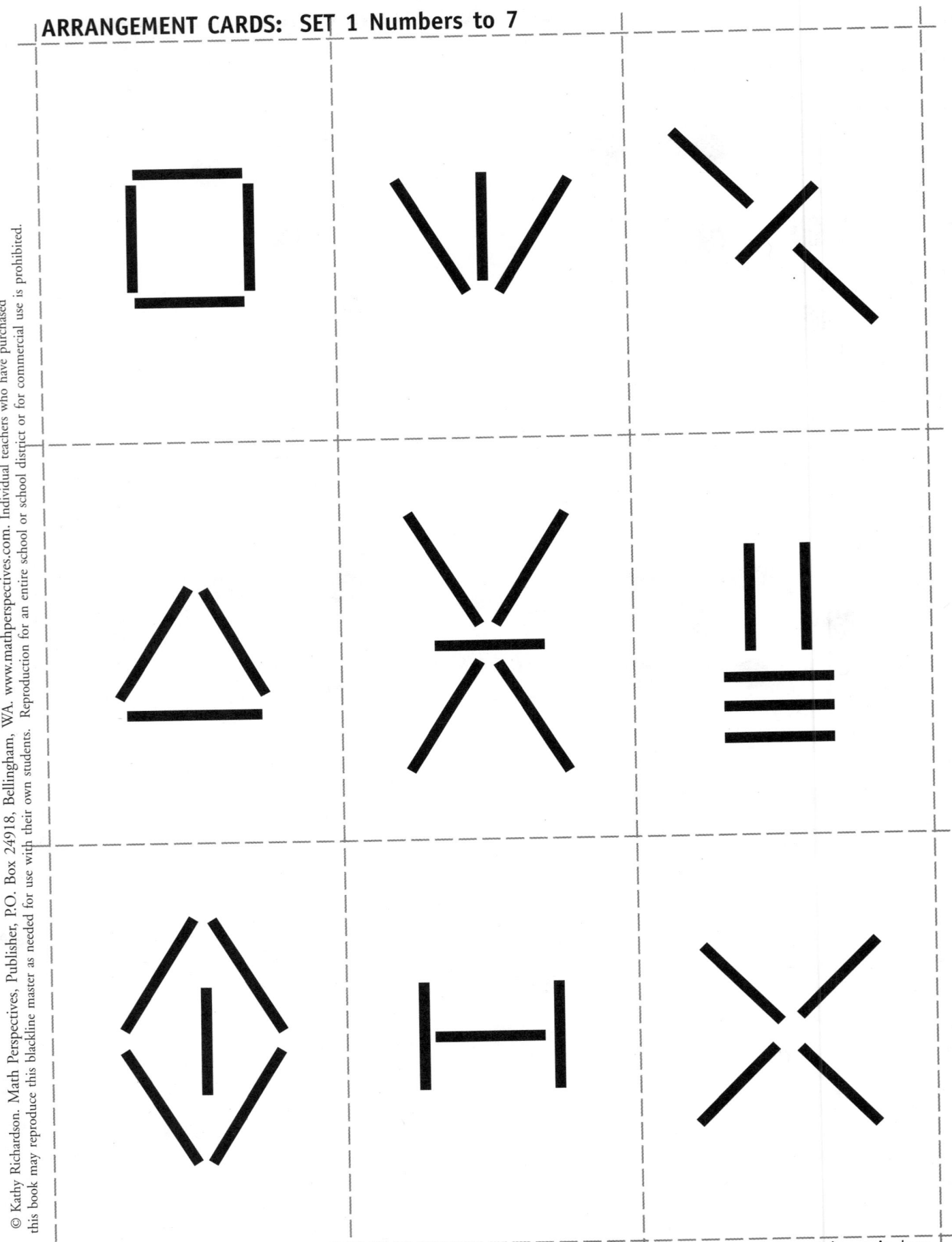

Copy onto card stock, cut on the dotted lines, and store complete set in a zip bag.
It can be helpful to use a different color for each set.

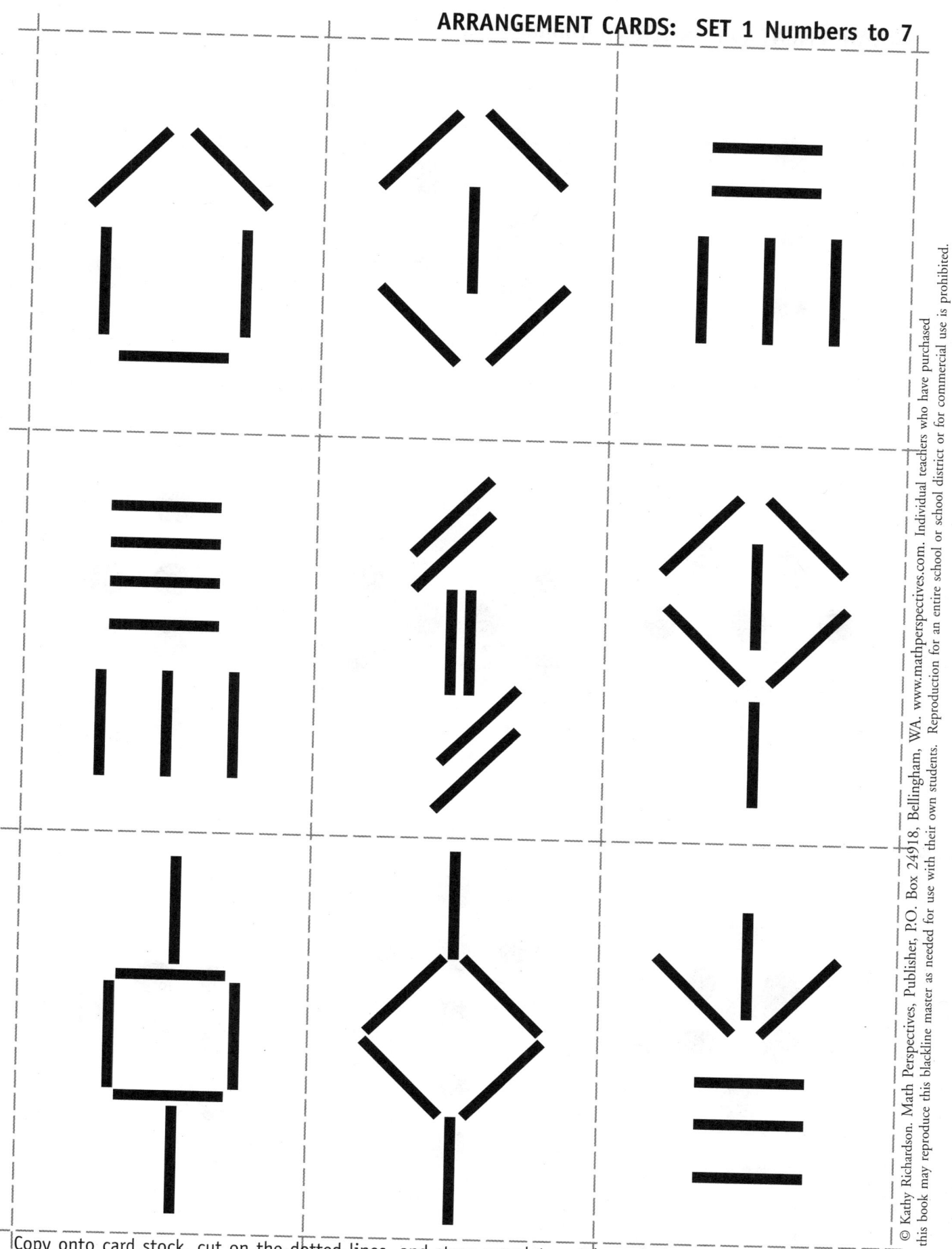

Copy onto card stock, cut on the dotted lines, and store complete set in a zip bag.
It can be helpful to use a different color for each set.

ARRANGEMENT CARDS: SET 1 Numbers to 7

Copy onto card stock, cut on the dotted lines, and store complete set in a zip bag.
It can be helpful to use a different color for each set.

Copy onto card stock, cut on the dotted lines, and store complete set in a zip bag.
It can be helpful to use a different color for each set.

BLM 4: 8

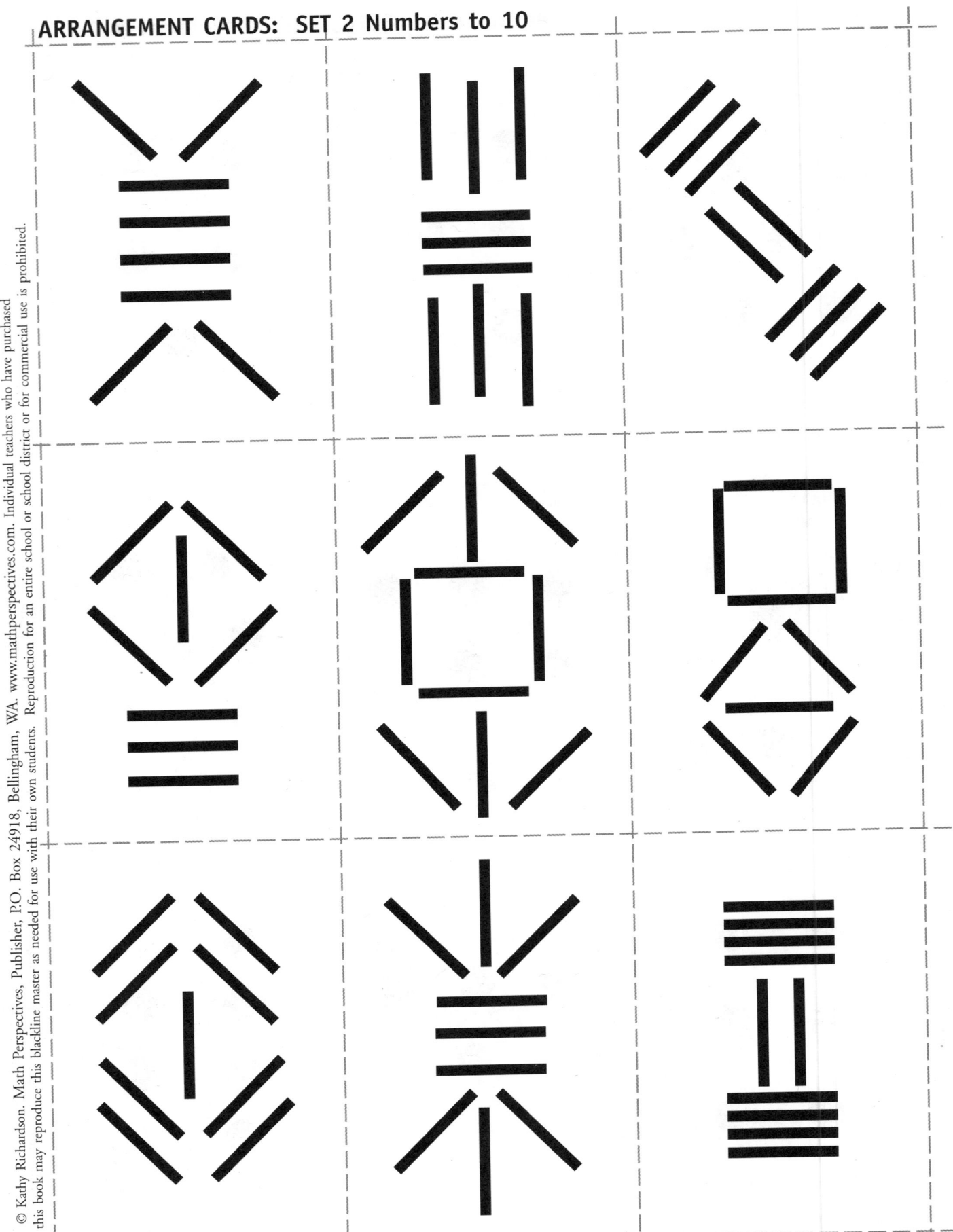

Copy onto card stock, cut on the dotted lines, and store complete set in a zip bag. It can be helpful to use a different color for each set.

Copy onto card stock, cut on the dotted lines, and store complete set in a zip bag.
It can be helpful to use a different color for each set.